The Palgrave Lacan Series

Jacques Lacan is one of the most important and influential thinkers of the 20th century. The reach of this influence continues to grow as we settle into the 21st century, the resonance of Lacan's thought arguably only beginning now to be properly felt, both in terms of its application to clinical matters and in its application to a range of human activities and interests. The Palgrave Lacan Series is a book series for the best new writing in the Lacanian field, giving voice to the leading writers of a new generation of Lacanian thought. The series will comprise original monographs and thematic, multi-authored collections. The books in the series will explore aspects of Lacan's theory from new perspectives and with original insights. There will be books focused on particular areas of or issues in clinical work. There will be books focused on applying Lacanian theory to areas and issues beyond the clinic, to matters of society, politics, the arts and culture. Each book, whatever its particular concern, will work to expand our understanding of Lacan's theory and its value in the 21st century.

More information about this series at
http://www.palgrave.com/gp/series/15116

Calum Neill
Editor

Lacanian Perspectives on Blade Runner 2049

palgrave
macmillan

Editor
Calum Neill
School of Applied Sciences
Edinburgh Napier University
Edinburgh, UK

The Palgrave Lacan Series
ISBN 978-3-030-56753-8 ISBN 978-3-030-56754-5 (eBook)
https://doi.org/10.1007/978-3-030-56754-5

Cover illustration: Pictorial Press Ltd/Alamy Stock Photo

This Palgrave Macmillan imprint is published by the registered company Springer Nature Switzerland AG
The registered company address is: Gewerbestrasse 11, 6330 Cham, Switzerland

For Claire,
with one more kiss, dear.

Contents

Notes on Contributors

Alexander Bove is an associate professor of English at Pacific University, where he teaches courses in nineteenth-century British literature, critical theory, and film theory. His articles have appeared in several journals, including *LFQ: Literature/Film Quarterly*, *Mediations: Journal of the Marxist Literary Group*, *ELH: English Literary History*, and *V21 Collective*. He is author of the book *Spectral Dickens: The Uncanny Forms of Novelistic Characterization*, forthcoming on Manchester University Press, and is currently at work on a book entitled *Extimate Materialism that Explores the Relation Between Film* and the novel through critical theories of comedy, the uncanny, and characterization.

Daniel Bristow is a psychoanalyst and writer, and co-creator of the Everyday Analysis project. He has published widely on Lacanian psychoanalysis, critical theory, and politics, and is author of *Joyce and Lacan: Reading, Writing, and Psychoanalysis*, and *2001: A Space Odyssey and Lacanian Psychoanalytic Theory*, which is also in the Palgrave Lacan Series.

Scott Contreras-Koterbay received his Ph.D. from the University of St Andrews and is a professor in both the Department of Art & Design and the Department of Philosophy & Humanities at East Tennessee State University, where he teaches aesthetics, the aesthetics of technology, artistic identity and contemporary art history as well as being the Director of the Bert C. Bach Fine & Performing Arts Scholars program in the Honors College. He is the author of *The Potential Role of Art in Kierkegaard's Description of the Individual* (2004) and co-author with Łukasz Mirocha of *The New Aesthetic and Art: Constellations of the Postdigital* (2016).

Matthew Flisfeder is associate professor of Rhetoric and Communications at The University of Winnipeg, Canada. He is the author of *Algorithmic Desire: Toward a New Structuralist Theory of Social Media* (2021), *Postmodern Theory and Blade Runner* (2017), and *The Symbolic, The Sublime, and Slavoj Žižek's Theory of Film* (2012), and co-editor of *Žižek and Media Studies: A Reader*.

Sheila Kunkle is associate professor of Individualized Studies at Metropolitan State University. She has published numerous articles on psychoanalysis and culture, and contributed chapters on the psychoanalysis of film to Psychoanalyzing Cinema (Jan Jagodzinski, ed. 2012) and Lars Von Tier's Women (Rex Butler and David Denny, eds., 2017), as well as edited the collection, *Cinematic Cuts: Theorizing Film Endings* (SUNY Press, 2016).

Todd McGowan teaches theory and film at the University of Vermont. He is the author of *Universality and Identity Politics*, *Emancipation After Hegel*, *Only a Joke Can Save Us: A Theory of Comedy*, *Capitalism and Desire*, and other works. He is the co-editor of the *Diaeresis* series with Slavoj Žižek and Adrian Johnston at Northwestern University Press and editor of the *Film Theory in Practice* series at Bloomsbury.

Isabel Millar recently received her Ph.D. in psychoanalysis and philosophy from Kingston University. Her thesis is entitled "The Psychoanalysis of Artificial Intelligence." Her work has appeared in *Stillpoint Magazine*, *Psychoanalytische Perspectieven*, *Vestigia*, *JCFAR journal*, and forthcoming publications for the *Courtauld Institute of Art* and *Precog Magazine*. She is also a screenwriter and psychoanalytic script consultant for film and TV.

Calum Neill is associate professor of Psychoanalysis & Cultural Theory at Edinburgh Napier University, Scotland, and Director of Lacan in Scotland. He has written a number of monographs, including *Without Ground: Lacanian Ethics and the Assumption of Subjectivity* (2011) and *Jacques Lacan: The Basics* (2017). He is the co-editor of both the *Palgrave Lacan Series* the three volume guide *Reading Lacan's Ecrits* (2018–2021).

Timothy Richardson is associate professor of English at the University of Texas at Arlington, where he teaches courses in theory, media, and writing. He is the author of *Contingency, Immanence, and the Subject of Rhetoric* (2013) as well as articles in such journals as *Kairos*, *Pre/Text*, and *Enculturation.*

Ben Tyrer is a lecturer in film theory at Middlesex University. He is the author of *Out of the Past: Lacan and Film Noir* (Palgrave, 2016) and co-editor of *Psychoanalysis and the Unrepresentable and Femininity and Psychoanalysis* (both Routledge, 2016 and 2019). He is a member of the editorial board of the *Film-Philosophy* journal and co-coordinator of the Psychoanalysis in Our Time research network.

Slavoj Žižek is senior researcher at the Department of Philosophy, University of Ljubljana, Slovenia; co-director of the Institute for Humanities, Birkbeck School of Law, London; and visiting professor at the Kyung Hee University, Seoul. Fields of work: philosophy, political theory, critique of ideology. Latest publications: *HEGEL IN A WIRED BRAIN* (London 2020), *PANDEMIC* (New York 2020).

List of Figures

1

From Voight-Kampff to Baseline Test: By Way of an Introduction

Calum Neill

The original Blade Runner film, set in November 2019, opens with the now iconic scene of Leon, a replicant, undergoing what appears to be a psychological association-reaction test. He complains of getting nervous when he takes tests but is told not to worry. "You're in a desert," Holden, the test administrator, tells him, "Walking along in the sand, when all of a sudden you look down…"

"What one?" Leon interrupts. He is told it doesn't matter, that it is completely hypothetical. But he persists, asking how he would have come to be there.

"Maybe you're fed up." Holden tells him, adding some emotional flavour. "Maybe you want to be by yourself. Who knows?" Then he continues with the script. "You look down and see a tortoise, Leon. It's crawling toward you…"

"Tortoise? What's that?"

C. Neill (✉)
School of Applied Sciences, Edinburgh Napier University, Edinburgh, UK
e-mail: c.neill@napier.ac.uk

C. Neill (ed.), *Lacanian Perspectives on Blade Runner 2049*,
The Palgrave Lacan Series,
https://doi.org/10.1007/978-3-030-56754-5_1

Holden asks him if he has seen a turtle and tells him it is the same thing. When Leon says that he's never actually seen a turtle, Holden begins to show irritation. Leon picks up on this and reassures Holden, saying "But I understand what you mean."

Holden resumes, "You reach down and you flip the tortoise over on its back, Leon."

But Leon is having trouble focusing. "Do you make up these questions, Mr. Holden? Or do they write them down for you?"

"The tortoise lays on its back," Holden continues, the audio track reverberating, presumably allowing us, the viewer, to enter into Leon's disorientation, "its belly baking in the hot sun, beating its legs, trying to turn itself over. But it can't. Not without your help. But you're not helping."

"What do you mean, I'm not helping?"

"I mean, you're not helping. Why is that, Leon?"

Leon's distress is now quite clear. Holden changes his tone and attempts to reassure him.

"They're just questions, Leon. In answer to your query, they're written down for me. It's a test, designed to provoke an emotional response." It appears to have succeeded. "Shall we continue?" Holden asks. He continues.

"Describe in single words only the good things that come into your mind, about your mother …"

"My mother?"

"Yeah."

"Let me tell you about my mother," replies Leon, leaning forward, his hands under the table. And he shoots Holden.

The Voight-Kampff test is, as Holden says, designed to provoke an emotional reaction. The apparatus Holden unfolds at the outset of the test, functioning a little like a lie-detector, measures physiological changes, with a particular emphasis on eye movement. The logic of the test appears to be rooted in emotion. Replicants, the humanoids manufactured by the Tyrell Corporation, emerge as fully formed adults. They have no childhood and therefore no childhood memory. They do, however, appear to be capable of desire, with the suggestion, then, of some kind of emotional attachment. The test, in the small samples we are shown of it, seems to operate on the basis of provoking the exposure of the gap between the awareness of the appropriacy of emotion and the lack of such appropriate emotion. When Holden describes the overturned tortoise, Leon appears to know that a reaction is expected of him and yet he doesn't know, or doesn't feel, what this reaction is.

The dénouement of the scene, with the invocation of the mother, seems pertinent in a psychoanalytic context. Where the tortoise merely provokes discomfort, the mention of his mother provokes a strong, violent, or murderous, reaction. Except Leon doesn't or didn't have a mother. It is plausible that Leon's reaction is nothing at all to do with the specific content of Holden's questions and is simply a pre-emptive reaction to the obvious point that he is about to be found out as being a replicant. And yet, the content cannot be ignored.

The test can be understood to operate on a logic of difference. The presence of an appropriate reaction—whatever that might be—would, presumably, indicate a likelihood that the subject is not a replicant. We might assume then, that the absence of an appropriate reaction would indicate that the subject is a replicant. However, this is not the case. It is the anticipation of the absence of an appropriate emotional reaction on the part of the subject themselves which appears to be the true point of confirmation. The test centres on the subject's own knowledge of their status, whether this knowledge is consciously known or not. It is not, however, a knowledge of what is but, rather, a knowledge of what might not be.

This point of anticipation exposes something crucial of the Cartesian core of the original film. The three central characters each occupy a particular stance towards the question of their knowledge of their own essence. Roy Baty, the leader of the rogue gang of replicants knows that he is a replicant. Rachel, a prototype of a newer model of replicant, appears to know that she is a replicant but struggles to acknowledge this knowledge. She knows but does not believe (Neill, 2018; 218). Deckart, the Blade Runner, we might, then assume, knows he is not a replicant. A key driver of the film, however, is the uncertainty of this knowledge. We, the spectator, are led to doubt the veracity of Deckart's knowledge, without this doubt ever settling into a new certainty. Forty years after the original film's release, through various allow is the anticipation of certain termination that the original Voight-Kampff test triggers. In the replicants' world of *2049*, there is nothing to anticipate. Not only has castration has always already occurred, as, of course, is true for the replicants of the original *Blade Runner*, but, moreover, the acceptance of castration as an irreversible fact is established from the off.

The relationship between castration and language is paramount here and underscores the significance of the shift from Voight-Kampff Test to Baseline Test. Against the commonplace notion that language is a vessel for the transmission of preformed ideas, Lacan helps us to appreciate that something like the opposite is true. For Lacan it is not that I, as agent, select the appropriate pieces of language to convey my already existing thoughts. Rather, for Lacan, I am myself produced as an incomplete entity on the basis of a language which precedes and envelops me. Language exists as an impersonal, temporally infinite and always incomplete chain of elements or signifiers. These signifiers, which have no meaning, articulate to each other and, in the process, articulate the subject. The signifiers are, quite literally, cells interlinked. In terms of signification, they are a black nothingness, a conceptual emptiness, an imposing lack, wanting to be filled. This is the castration of language. Without it we are lost to the nothingness. With it, which is the only meaningful possibility at all, we are lost to ourselves.

To speak of castration here, then, is to speak also of lack and of desire. It is to speak, then, of the replicant as a subject. But as what kind of subject? This is a question which is picked up in various ways through the different chapters of this book. The already and irrevocably castrated subject who knows their castration is beyond both question and repair, is a subject without hope. It is not, however, a subject without desire. Beyond hope, the subject is locked in place. But without desire, the subject would not be a subject. Thus, while the hypercapitalist context of the *Blade Runner* cinematic universe invites comparisons of the replicant to the slave or the disenfranchised labour class, more than anything, it invites comparison with the quotidian subject of capital. We might say that it is precisely this combination of hopelessness and desire which not only defines the subject of capitalism but facilitates capitalism's continuation.

The irrevocably castrated subject is what might conventionally be called the male subject. As we know from Lacan's formulae of sexuation, it is not that women are not castrated, but rather that they are 'not all' castrated, with all the ambiguity that that 'not all' invites (Lacan, 1999; 101). In the original *Blade Runner*, there are few female characters and, with one exception, they are presented as pleasure or entertainment models. They exist as the plaything or for the titillation of the male characters. The exception, Rachel, does not veer too far from this template. She is not explicitly created exclusively for pleasure but, nonetheless, she is an object of exchange for the male characters.

Representations of femininity abound in *2049*. They are magnified, in terms of their representative scope, their ontological status and their sheer volume. The characters occupy a greater variety of roles, from police chief to prostitute to rebel leader to corporate enforcer to holographic housemaid, although only one of these is human. The landscape of the city is surfeit with images of the female body, emphatically in the singular, as the same body appears to be repeated in various sizes. Echoing this, when K ventures out Los Angeles in his quest to find Deckard, he is welcomed to Las Vegas by towering statues of the female form. The plot itself, while ostensibly focused on a male character, sees him move from one woman to another, pursued by one, seeking another. Where the female characters in *Blade Runner* are there to be desired, the basic plot of *2049* seeks to foreground a different role for woman; child birth. But where this shift folds back on itself is in that, even in this move towards an 'essential femininity', it is still presented as being for the male. Miraculously, Rachel, the exceptional replicant from *Blade Runner*, has had a child. K. is tasked with tracking down the child but, erroneously, convinces himself that the child is himself. It is only at the end of the film, when we discover that K. had been but a diversion, that it emerges that Rachel's true offspring is a woman. This liminal woman, Ana, who can be categorised neither as a replicant nor a human, is the creator of the pre-formation memories which are subsequently programmed into replicants. Powerful and gifted as she may be, she is effectively another slave of Wallace.

Femininity and capitalism emerge then, not simply as key thematic talking points of the film but, much more than this, they can be seen to be the entwined pillars of the film, each supporting the other and each demanding an excavation of the other in order to further an understanding of how we consume the film, consciously or not. To be a human, to be, in Lacanian terms, a subject, is to be a sexed subject and to be human now, to be a subject of capitalism, is to be a sexed subject of capitalism. One's sexual position is not then an adjunct to one's place in society. It defines it. The peculiar binary that late-capitalism offers is sexual but it is not a binary of distinct and pregiven positions. It is a binary of exclusion.

It is perhaps not accidental then that, not only are all of the female characters in the first *Blade Runner* replicants, but the majority of the replicants are female. Even Roy Baty, hypermasculine in some regards, is presented, in terms of Lacan's formulae of sexuation, as logically female. By the time of *2049*, this position has shifted, at least in emphasis. The cast of replicants still seems to be female, although now many, like Luv, are logically male. The central replicant, however, is now resolutely male, just as resolutely male as Deckard in the first film, a coincidence which perhaps lures the viewer into the trap of assuming K to be Deckard's son and propels us back to the question of the relationship between human and non-human. This conceptual passage has always been a key element in theoretical discussions of the original *Blade Runner*. The film raises questions regarding what it means to be a human and how we relate to that which is non-human, particularly when the non-human's appearance and character appear to offer the opportunity of misrecognition with our own.

Blade Runner 2049 can be understood as developing, problematising and extending many of these questions. Key motifs of sex, social order and the rule of capital, which are clearly evident in the first film, are fruitfully extended in the sequel. The intersection of these concerns with the core question of being is, then, similarly extended, and psychoanalytic theory can help us both grapple with and further problematise the issues raised. Such grappling and problematisation are the focus of the first essay following this one; Ben Tryer's 'Do Filminds Dream of Celluloid Sheep?'. Exploring many of the conceptual extensions, and short-circuits, in the relationship between the original and the sequel, Tryer masterfully explores the central concerns of being and knowledge, showing how even a less than favourable reading of *2049* can help us to move beyond the Cartesian coordinates of the original *Blade Runner* and open up an understanding of film, and *this* film, as operating on and with the unconscious.

The place of the unconscious is also a central question for Slavoj Žižek as he seeks to explore how we understand the unconscious as we accelerate into late techno-capitalism and how the film's depiction of the conceptual relation of human to the nonhuman opens up this vital question and allows us to posit the replicant as an adumbration of the worker of the future. This reading, of the replicant as a foreshadowing of the worker or, we might say, the subject of late capitalism, is continued in many of the essays which follow.

Focusing on the inherent tension capitalism presents between the unfettered market and the need for social control, Todd McGowan argues that this tension offers up a revolutionary potential. Harking back to the Cartesian subject who establishes his own existence through the certainty of the fact of doubting, McGowan deftly argues that the very cynical distance that capitalism encourages is its own protection. It is only as that cynical distance diminishes and something like belief emerges that doubt again becomes a possibility. This potential for social revolution is further developed in Daniel Bristow's essay where he explores the place of the *sinthome* and how this might articulate with utopian urges to resist the fatal flaw of the utopic itself; the contradiction of its closing down. For Bristow, the *sinthome* is precisely what might allow the utopic to be maintained as unobtained, allowing persistent productive potential. Such, albeit cautious, optimism is offset by Timothy Richardson's essay which warns of the recuperative nature of fantasy in late capitalism. Against a commonplace understanding of fantasy as an escape from real life, Richardson emphasises a Lacanian understanding of fantasy as that which allows the possibility of subjective life. The danger in the system of capitalism is that, no matter how individually construed, fantasy is forged on the basis of social discourses which tend to be controlled by capital. Thus, the very fantasy of escape, the awakening of constant K.'s inconsistency, is the lure that returns us to the market. As the object of desire, for which fantasy provides a scene, is never it, so desire continues. While such desire may, as McGowan and

Bristow suggest, open up the difficult possibility of change, it can just as easily be mobilised to maintain the circulation of capital.

Matthew Flisfeder, Alex Bove and Scott Koterbay all pick up on themes of anxiety in relation to the posthuman and the perceived threat of technology, drawing our attention to different aspects of the film and how it can be read in such a way as to highlight different and related concerns in late capitalism. Drawing on the theory of Object Oriented Ontology, Bove argues that *2049* unsettles the very distinction that we might want to hold between human and object to disconcerting effect. Continuing this engagement with OOO and the contemporary anxiety over automatism, Flisfeder argues that the film opens a discourse on the manners in which we experience fantasy, desire and enjoyment in late capitalism. Such a discourse allows us to appreciate the shift from positing the subjugated other as object to perceiving them, however anxiety provoking it might be, as subjects. These anxieties are explored further in Koterbay's essay where he argues that technological progress and advances in artificial intelligence are increasingly overshadowing the human being in a manner that poses what we might call an existential threat. As such artificial intelligences come to operate as subjects with their own desires and, indeed, anxieties, we begin to see the mirror reflect our own imminent obsolescence.

Maintaining a focus on fantasy and desire, Isabel Millar turns our attention to the place of the woman, both in the film and in late capitalism. As noted above, the relative proliferation of female roles, in relation to the original *Blade Runner* film, and the various manners in which they provide structure for the otherwise hollowness of K.'s subjectivity, allows *2049* to be read, perhaps against itself, as a presentation of the failure of masculinity as it struggles to disavow the feminine as something irrecuperable to its own form. This notion of the female characters as sustaining the possibility of subjectivity is picked up upon by Sheila Kunkle, although she warns of the fragility of this position in the face of patriarchy's tendency to recuperate, drawing our attention to the film's ending which, once again, places the male at the centre of the narrative.

References

Blade Runner. Dir. Ridley Scott. Warner Brothers, 1982. Film.
Nabakov, V. (1962) *Pale Fire*. London. Penguin.
Neill, C. (2018) Do Electric Sheep Dream of Androids? On the Place of Fantasy in the Consideration of the Nonhuman. In G. Basu Thakur and J. M. Dickstein (eds.) *Lacan and the Nonhuman*. London: Palgrave Macmillan.
Lacan, J. (1999) *The Seminar of Jacques Lacan Book XX: Encore—On Feminine Sexuality, the Limits of Love and Knowledge*. London: W.W. Norton & Company.

2

Do Filminds Dream of Celluloid Sheep? Lacan, Filmosophy and *Blade Runner 2049*

Ben Tyrer

Introduction: Psychoanalysis and Avoidance

At the origin of the "Blade Runner" universe is a question: *Do androids dream of electric sheep?* It brings together two key phrases from Dick's novel: Deckard's question, "Do androids dream?" and the figure of the "electric sheep". At first face, this curious title already seems to put us on a road—as Freud once said—to knowledge of the unconscious: or rather, here it puts the unconscious in *question.* Do androids dream? Do they dream of sheep? Are they electric? In a properly psychoanalytic context, these latter two should be subordinated to the main inquiry: the "content" of such dreams being secondary to the very fact of dreaming itself, which suggests a certain structure of *desiring*. If androids *dream,* then we could say that *the unconscious is in play.* However, closer inspection of Dick's novel reveals that what is meant by "dream" here is closer to

B. Tyrer (✉)
Middlesex University, London, UK
e-mail: b.tyrer@mdx.ac.uk

C. Neill (ed.), *Lacanian Perspectives on Blade Runner 2049*,
The Palgrave Lacan Series,
https://doi.org/10.1007/978-3-030-56754-5_2

"aspire" and the "electric sheep" is not the manifest content of a nocturnal psychosis but the robotic replica of the living *Ovis ares*: a highly-coveted object in the world of Rick Deckard. As much as it is an interrogation of identity and posthumanism, the novel is thus a story of consumerism and commodity fetishism (for example, instead of eloping with his replicant partner, Deckard ends the novel still very much human, in a troubled marriage and tending to a valuable "electric toad" that he has recovered from the desert). Rather than a journey along Freud's via *regia*, then, this in fact aligns the work with the historical uptake of psychoanalysis in the USA as "ego psychology". Seeing it as based in notions of strengthening the ego, Lacan considered the practice—as Jane Gallop observes—as "a deformation of psychoanalysis peculiarly suited to American values, giving the American people what they want" (1987: 57). Ego psychology, in Lacan's estimation was the handmaid of the *dream of electric sheep*, facilitating the individual's pursuit of happiness (via commodities) rather than the encounter with unconscious desire.

Similarly, while shifting registers quite radically, we can observe another avoidance of the unconscious in the contemporary field of Film Studies. Not unlike the Wallace Corporation of *Blade Runner 2049* (Villeneuve, 2017), Daniel Frampton's *Filmosophy* project posits (or seeks to create) a new form of (artificial) thinking and feeling by conceiving of cinema itself as a kind of mind: a *filmind*. He presents a partial overview of film theory and philosophy that has already attempted to consider film in relation to ideas of thinking and mental processes, and then proposes a novel form of "film-thinking" as a means of conceptualising the ways in which the medium creates meanings and affects. Constituting, as I will explain, a key example of the movement *away* from psychoanalysis in Film Studies, *Filmosophy* figures the thinking of Freud and Lacan (and the film theorists they have inspired) only at its margins: even where it would seem to come into close proximity to notions such as *dream* and *fantasy*. Psychoanalysis seems to haunt *Filmosophy* at various points, but this spectral presence becomes further distorted by the strange disavowals of the unconscious that Frampton expresses. And nowhere is this clearer than in the single explicit reference to Lacan (or at least "Lacanians") for whom, Frampton asserts, "the subconscious [sic] is structured like a language" (2006: 150). Frampton is clearly not *unaware* of psychoanalysis, and even offers a lengthy footnote that aims to spell out some of the difference between "unconscious" and "subconscious" in his estimation (2006: 230n6), but such a direct error in relating perhaps Lacan's most famous dictum can only be read symptomatically.

As Todd McGowan observes, "replac[ing] the term 'unconscious' with 'subconscious,' as often happens today (...) minimizes the alien status of the unconscious and serves to domesticate it" (2015: 19). Frampton's *slip* here attempts to disempower the psychoanalytic project, which would seek to emphasise the significance of the slip itself as a means of approaching the alterity of another mode of thinking: in the unconscious. Reflective, as I've said, of a more general retreat from psychoanalysis, the filmind of *Filmosophy* needs to be re-encountered *with* the unconscious in order to consider the cinema experience once more. In what follows, then, I will ask whether *filminds dream of celluloid sheep*: which is to say, contrary to Dick's formulation, whether the encounter with the "Blade Runner" universe and psychoanalysis would insist upon a wholesale rethinking of Frampton's project (or perhaps just a slight Lacanian adjustment). In order to offer an answer, I aim to chart some of the ways in which film and philosophy alike have attempted to avoid, ignore or otherwise deny the alterity of the unconscious through an approach to the questions of the subject and repetition in *Blade Runner 2049* via both filmosophy and Lacan.

Unthinking Film-Thinking

In order to establish the conceptual coordinates of this chapter, it will be necessary first of all—continuing along this line of inquiry into (missed) encounters with the psychoanalysis—for me to take a brief excursus through the field of film-philosophy to help further clarify the methodological framework with which I will approach *Blade Runner 2049*. This begins with Frampton's *Filmosophy*, which introduces itself (on the book's front cover) as "a manifesto for a radically new way of understanding cinema". *In nuce*, Frampton's claim—drawing on Deleuze and Vivian Sobchack—is that we should conceive of the film experience as an encounter with a *mind*, but one utterly different from the human mind: a "filmind" (2006: 6). The film is thus "thought" by this filmind, which creates a world that should be understood as *intended*—in the philosophical-phenomenological sense—by the filmind in its capacity to produce and organise images and sounds. This "film-thinking" (2006: 6) has the benefit, Frampton suggests, of providing the "filmgoer" with a *poetic* means of contemplating and experiencing the film on its own terms, free—for example—from what he considers to be the deadening language of neo-formalism and the suffocating methods of technicist film analysis. Instead of "tracking shots" and "graphic matches", the film "thinks" its world in terms of a forward or sideways movement or even "feels" it in terms of intensities of colour and shape. This is not, to be clear, an ontological claim: Frampton is not asserting that film literally *is* a mind but an articulation of the film experience in terms of the concepts of mind and thinking.

Frampton's project is situated, if not always explicitly, in *opposition* to psychoanalytic thinking on film. As I have argued elsewhere, the recent turn towards philosophy in the field of Film Studies has been predicated on a rejection of a particular version of Lacan.[1] However, such engagements with psychoanalysis are often severely limited, being based on a view of only certain very specific elements of Lacan's work, while not taking on board both his wholesale re-evaluation of Freud's legacy and the various reversals and developments in his own thinking of the unconscious. This refusal reverberates throughout Frampton's version of the history of film theory, which provides only a *partial* take on thinking cinema and mind together that does not engage with psychoanalysis in any meaningful way. *Filmosophy*, however, continually evokes psychoanalytic concepts and processes without acknowledging them as such. Frampton states consistently that the concept of filmind should be opposed to any understanding of the human mind: "filmosophy does not make a direct analogy between human thought and film, because film is simply different to our ways of thinking and perceiving" (2006: 7). He explains, "It is called 'filmind' because, simply, it is not a human mind. It is another kind of mind, its own mind, a new mind" (2006: 73). If it is the case, as Frampton claims, that "To find objectifications of the mental—to theorise an external consciousness, to see a mirror to the soul—is to drag film down to our cognitive-rational level. Film is more than this" (2006: 26), then a psychoanalytic theorist would agree. For example, the cognitive, mind-as-computer model of film experience is severely limiting: but then, as we shall see, without a place for the unconscious so is what Frampton offers in its stead. Indeed, where he claims, "film-thinking is not mappable by the terms of human thinking" (2006: 151), I would add that neither is psychoanalytic thinking, if we conceive of "human thinking" as something that excludes the unconscious. In fact, the model of "human subjectivity" that Frampton rejects as a basis for filmosophy is precisely the model that psychoanalysis rejects too.

While filmosophy refuses an analogy between filmind and human mind, Frampton's understanding of this "human mind" is limited—I'd argue—to what Mladen Dolar identifies as the "cogito" of modern philosophy (1998: 11). In a discussion of film-thinking and phenomenology, Frampton notes that Sobchack's approach allows her to posit film as "think[ing] its objects" as an intentional, embodied subject, it thus becoming an "animate, conscious other" (2006: 44). However, he reproaches her for what he sees as the anthropomorphism that phenomenology implies: reducing film to a "human-like" expression and vision, which "'makes visible' acts of consciousness" (2006: 46). Frampton rejects the notion that film has such human qualities, but does not interrogate the underlying assumptions about the subject determining his own analysis. His critique of Sobchack reveals his project's implicit theory of the "human subject" as philosophy's "self-transparent subjectivity" (Dolar 1998: 12). Frampton observes, apropos of Sobchack, that "if we were to call [filmind] 'subjective', it would be a different subjectivity to that which we express and hold" (2006: 42): a formulation that seems to imply an autonomous individual somehow *separate from* their subjectivity (we hold *it* rather than it holding *us*, for example). Frampton's charge of anthropomorphism thus stands only if we accept the coincidence of the "human" and the "subject". However, Lacan's crucial insight is his insistence on the radically *anti-human* subject: one that cannot be reduced to or understood in terms of the (liberal humanist) self-transparent individual of filmosophy (and, as Dolar notes, the last three hundred years of Western philosophy in general [1998: 11]).

This is where the philosophical import of Lacan's project—as a science of the subject—becomes apparent. The Lacanian subject is not the *ego* and—for that matter—neither is the psychoanalytic ego equivalent to the subject of modern philosophy (*qua* "cogito"); it is not an autonomous agent but an illusory wholeness formed through alienating identification with the image of the other. Nor is the Lacanian subject equivalent to modern philosophy's "cogito", but then—for Lacan—*this* "cogito" is not equivalent to Descartes' *cogito*, the I of the "I think", *either* (cf. Dolar 1998). Lacan's resituation of the cogito is largely beyond the scope of this chapter—although Žižek elaborates this idea in relation to *Blade Runner* (Scott, 1982) in particular in *Tarrying with the Negative* (see 1993: 12, and below)—but for now we can observe that when Frampton rejects anthropomorphism for filmosophy, psychoanalysis would be in agreement: the Lacanian "subject"—as the subject of the unconscious—is not an anthropomorphic model, either. Frampton's understanding of the individual holds *no place* for the unconscious, just as his vision for filmosophy hold *no place* for psychoanalysis: but in both instances *the Lacanian subject haunts the scene.*

As Lacan states, "the unconscious is the Other's discourse" or "the unconscious is the discourse of the Other" [*discours de l'Autre*] (2007: 10; 1977: 45). Lacan moves understanding of the Freudian unconscious away from an *individual* unconscious grounded in speech to a *transindividual* unconscious related to the Symbolic. He states that "this so-called internal monologue [the individual unconscious] is entirely continuous with the external dialogue, and indeed this is why we can say that the unconscious is also the discourse of the Other" (1993: 113). The Other is thus constituted not as a person but as a *place*—"it must first of all be considered a locus" (1993: 274)—and the register of the "truth" of the subject, moreover, is "situated somewhere else altogether: at the very foundation of intersubjectivity" (2007: 13). It is the locus from which the subject "receives his own message in an inverted form" (2007: 30); whence the symbolic Other "returns" the message in its *true* form, as formations of the unconscious. In *this* context, therefore, where Frampton asserts that film creates its world "not from a 'point of view', but from a realm, a no-place, that still gives us some things and not others" (2006: 38), I'd be tempted to return his message—in its true, inverted form—restating it thus: the filmind produces sounds and images from an-Other realm, the "no-place" of the unconscious. If, for Frampton, film doesn't show us "human thinking", then *neither does psychoanalysis*. If "human thinking" is the rational, conscious agency implied by Frampton and assumed by, for example, cognitive philosophers and film theorists, then psychoanalysis stands as directly opposed to it as filmosophy does. And where Frampton relates the filmind to Deleuze's "spiritual automaton" as "a mode of thought that is alien and outside to this normal thinking" (2006: 65), I would assert a direct connection to the thinking of Lacanian unconscious once more.

Indeed, as Žižek observes, in making "the radical externality of the Other the place where the truth of the subject is articulated" (1992: 76), Lacan shows the unconscious to be an extrasubjective locus rather than an intrasubjective agency. The unconscious is realised in the Other and if it seems as though this is "interior" it is because its true relation to the subject is blocked by the Imaginary. Lacan insists that, "[t]he fact that the symbolic is located outside of man is the very notion of the unconscious" (2007: 392). This Other is the Symbolic order in its individual relation to the subject, and its radical alterity is therefore the otherness of the unconscious. As *subject of the unconscious*, the Lacanian subject is thus ex-centric and heteronomous to itself. The unconscious is "external" to the subject, yet it seems to meet the subject at an uncannily intimate place. It is "the mode of absolute subjectivity, insofar as Freud truly discovered it in its radical eccentricity" (2007: 347). And this, moreover, can be connected to Lacan's neologism, "extimacy" [*extimité*], which emphasised the feeling of "exterior intimacy" evoked, for instance, in the experience of the uncanny (1997: 139). The extimate is thus the position of the unconscious for the subject; it is the decentred-centre of the subject, what Miller calls the "intimate that is radically Other" (1994: 76). The unconscious is thus the "*Kern unseres Wesen*, 'the core of our being'" that is simultaneously a foreign body (Lacan 2007: 437).

The unconscious, then, is not a "humanlike" mind; it is alien and uniquely particular to its own dimensions, terms and processes, *just like the filmind*. Frampton presents filmosophy as a "*humanist* poetics" and asserts that "The concepts of filmind and film-thinking *naturally* give birth to *humanistic* terms of intention (belief, empathy, etc.)" (2006: 175, emphasis added), but we could consider it to be the humanist practice of an encounter with a "thinking" that is fundamentally *non-human*. Conversely, Lacanian psychoanalysis is—as Mark Fisher observes—"untethered from any naturalisation or sense of homeliness" (2016): an anti-humanist encounter with an anti-human thinking. The unconscious marks the inhuman core of the individual, the void that constitutes (and in which is constituted) the subject and renders it not-at-home, out-of-joint with itself: it cannot be bound to anything we would recognise as the humanist individual (inner essence, organic unity, etc.). The Lacanian subject is fundamentally *anti-human(ist)* because—as I have noted elsewhere—it is constituted not as a pre-given individual but at the intersection of various agencies, drives and structures that we identify with names such as "sexuality" and "the unconscious".[2] And here my discussion leads us back to *Blade Runner*: as Calum Neill notes in his reading of Deckard's predicament, "The final flaw in the temptation to anthropomorphise is not that we read human traits into the nonhuman [which would be Frampton's defence of the filmind] so much as we read them into the human" (Neill 2018: 222). Indeed, where Frampton's conception of the "human" (and consequently the non-human) refuses the anti-humanist practices of psychoanalytic insight, I'd argue that it also maps to a significant degree onto a Lacanian understanding. Frampton suggests that film-thinking *exceeds* our own thinking; our "impower", he insists, "lies in being unable to think images (or image-concepts) as clearly as film. Deleuze called this gap in our ability the unthought in thought" (2006: 166). I would call the "unthought in thought" simply the *unconscious*: the nonhuman gap asserted by Lacanian psychoanalysis in any conception of human identity.

Furthermore, in attempting to sum up the significance of the project of *Filmosophy*, Frampton states: "Being always conscious (always filming), [the filmind] has no 'subconscious', and relates more to pre-reflective consciousness" and emphasises once again that his project "reveal[s] the distance between film and human consciousness" (2006: 203). Here, I can only agree. From a Lacanian perspective, I would equally reject any notion of "subconscious"; while I would also assert the distance between my own approach to film and any understanding of "human consciousness" not grounded in Freud's insights. Frampton *does* hint that the filmind "is perhaps its own nestled nonconsciousness" (2006: 203), and—the metaphor tending towards depth psychology notwithstanding—I would claim once more the corollary here between cinema (and the filmind) and the properly psychoanalytic "nonconsciousness": the anti-human *unthinking* of the Lacanian *unconscious*.

A first working thesis, then, would be to assert the *speculative identity* of filmind and unconscious: seemingly opposed by Frampton but coinciding in notions of "thinking" distinct from human agency. The idea that the filmind *is* the unconscious is not an assertion that film "visualises" the unconscious directly, as might have been claimed by the Surrealists; rather, it is the claim that what Frampton identifies as a wholly new way of thinking—film-thinking as analogous in some way but distinct from "human thinking" and as that which transforms reality into a new world—has long been *thought* by psychoanalysis as the working of the unconscious. We might consider, then, that the encounter between film and filmgoer is akin to the encounter with the unconscious, as an alien thinking outside of ourselves. This would constitute a degree of synthesis between filmosophy and psychoanalytic film theory as formulated by McGowan, for instance, who claims that film presents an encounter for the spectator with their own traumatic desire by taking the *form* of the unconscious (2015: 10). McGowan suggests that films are created in order to appeal to the desire of the spectator, and relates this to the psychoanalytic understanding of dreamwork as a formal construction giving expression to unconscious desire: it thus becoming a sort of signifier of the Other's desire. But rather than as an attempt to engage the spectator *qua* Other, we could conceive of the film experience—via filmosophy—as something more akin to an encounter with the Other of the unconscious itself: the type of encounter that usually comes about only in everyday life through slips, jokes, etc. or in the specific setting of the clinic. An encounter with film(-thinking), then, is like an encounter with the "unthinking" of the unconscious: it obeys its own logic(s) and produces meanings and affects in its own ways. Dreams and filminds alike must be taken on their own terms and interpreted individually as an organised system.

Can I Offer You Thinking-Without-the-Unconscious, Instead?

One of the crucial questions Frampton poses in his filmosophy is: How does the filmind think about its characters, spaces and themes? What is its attitude towards them? (see 2006: 7). In brief, I could say here that the "Blade Runner" films in general "think" the desires of replicants through dreams, memories and fantasies: through *the stuff of the unconscious* itself. And in order to continue this encounter between filmind and unconscious, I will elaborate this answer in relation to *Blade Runner* and *Blade Runner 2049*: charting different modulations of subjectivity and articulations of the unconscious as presented in Lacanian psychoanalysis. As such, I'd argue that Deckard, Rachael and, moreover, K in *2049* embody *different* (and differing) modes of *negation* of the "gap" between a knowable self and the unknowable unconscious. In this context, I'd note that Neill is indeed correct to identify knowledge and belief as the crucial categories concerning the central "Blade Runner" question (Am I a replicant?) (2018: 318); however, I'd argue that they operate slightly differently from the way in which he suggests. Roy is, as Žižek recognises, "the subject who *knows* he is a replicant" (1993: 40; emphasis added)—he knows *in full*, closing the gap in the subject and standing for an impossible totality, a *psychotic* structure where there is no repression of knowledge, nothing is unknowable (not even mortality), and the unconscious is foreclosed—but knowledge plays different roles for both his Nexus kin and their descendants. Moreover, I would agree with Neill that Roy *knows and believes* he is a replicant, while Rachael *knows but does not believe* (cf. 2018: 318); the corollary of which would be that while Roy knows and believes he is not human, Rachael knows she is not human but doesn't believe it. However, where Neill suggests that "Deckard is the final piece in the puzzle" because "we are left unsure of [his] status" (2018: 318), I would counter that it is precisely because we *do* appreciate his status (as replicant) that Deckard offers a special case here as compared to Roy (or even Rachael). In the Director's/Final Cut at least, there is much less ambiguity regarding Deckard's identity: as Žižek observes, there is an overt parallel between the revelation of Rachael's artificial memories (predicated on a photograph) and the placing of the unicorn daydream during the subsequent scene in which the camera lingers on Deckard's own family memorabilia (1993: 11). This is, moreover, anticipated by prior the match cut from the owl to Deckard (as each looks from right to left) in Tyrell's office, and his question to Rachael: "*Artificial*?", and these rhyming shots are affirmed, quilted even, by the knowing smile with which Deckard acknowledges at the film's end the origami unicorn that seems to confirm what he, we and Gaff now know.

Following Neill's insistence on the epistemological coordinates of *Blade Runner*, then, in Deckard's case it becomes clear that what is at stake is a question of *unconscious knowledge* (or *knowledge in the unconscious*) specifically. It is the inaccessibility of the self to the self that makes it an I; or, as Žižek explains, "the paradox of self-consciousness is that *it is only possible against the background of its own impossibility*" (1993: 15). I am only conscious of myself to the extent that I am beyond my grasp as a Thing that thinks. In more classically Freudian terms, Žižek renders this as: "I am not simply identical to myself but have an unconscious, insofar as I am prevented from having direct access to the truth of my own being" (1993: 31). In relation to this, he cites Deckard's astonishment that Rachael doesn't know what she is: the implication being that, in fact, the *only way we can know anything about ourselves* is via some part, a "kernel of being", that remains unknown and unknowable. This inaccessibility of the Thing that thinks, Žižek designates at the "primordial repression" (1993: 15), that nucleus of the unconscious which attracts all other (secondarily) repressed material.

For Rachael, and certainly moreover for Deckard, that inaccessible kernel of being—that which is primordially repressed—is the unconscious truth that they are *replicants*. In order to maintain a *belief* in their own self-identity, this wound or lack of being must necessarily remain repressed: and can only be expressed *symptomatically* in, for example, Deckard's unicorn vision (an "unnatural" creature galloping through woods, like a replicant in the world) that Matthew Flisfeder astutely describes as the "repressed" content of the US Theatrical cut of the film (2017: 123), which then *returns* in the Director's Cut and beyond. The passage towards such knowledge is of course the project of a psychoanalysis, but Deckard's intervention with Rachael is too direct: attempting to force the brute fact of her identity in the place of coming to the unconscious through interpretation (as a means of shifting belief about one's self) and as a result, Rachael's belief remains obdurate. It is not enough simply to *know* the truth of my symptom; I have to *believe* it too in order fully to assume its identity.[3]

Žižek frames such a question in terms of a relation of transference: "It is this [inaccessible truth of my being] that I am looking for in others: what propels me to 'communicate' with them is the hope that I will receive from them the truth about myself, about my own desire" (1993: 31). The Voight-Kampff test short-circuits this relation—it is a sort of technocratic fantasy of an instant psychoanalysis machine that reveals one's inner truth at the press of a button—and this is why its subjects react so negatively (recall Leon's violent outburst and Rachael's continued denial). However, the bond between Deckard and Gaff comes closer to a therapeutic relationship: with the origami unicorn, Gaff communicates a signifier to Deckard that allows him to come to an understanding about himself. The folded paper is yet another "purloined letter" by which Deckard receives his own message from the Other: the *truth* he has always *known* about himself, which he finally accepts with a nod of the head. What I am claiming, then, is that at the outset of *Blade Runner*, Deckard and Rachael both operate under a "normal" or *neurotic* structure, predicated upon *repression.* The unconscious truth of their identities is constituted as inaccessible to them, and their trajectories through the film mark versions of a successful and an unsuccessful analysis, respectively.

Officer KD6-3.7 of *2049*, on the other hand, presents a different articulation of this relation to belief and self-knowledge that, in short, suggests the clinical structure of *perversion.* All "Nexus 9" models are created by Niander Wallace *knowing* that they are replicants. There can be no quests for self-knowledge or throes of existential angst if the replicant subject already *knows* precisely what it is. And in one sense, Wallace's vision here would seem to be that which I have already suggested is shared by Frampton and what Dolar calls the "cogito of modern philosophy": the totally self-transparent individual *without unconscious.* Structurally speaking, then, in all three instances we find a sort of *psychotic* subject—one without repression, where no part is inaccessible—but certainly lacking the violent outbursts of the "Nexus 6" series such as Roy. As such, K has been designed both to *know* and to *believe* that he is a replicant. There is no space for doubt or variation in Wallace's template. As the "baseline test" makes clear, obedience and consistency are carefully measured and controlled each time K has been in the field. Yet K *does* come to believe that he is something more or something different from all the other "skin jobs" in *2049*, and—as I've said—it is this interplay between knowledge and belief in Villeneuve's film that is crucial for our concerns here.

First, there is K's belief in the difference between human and replicant: which can be described as a basic natalist position. When he wavers in the face of Lt. Joshi's request to "erase" all trace of Rachael's child, including the child itself, he explains to her:

K: I've never retired something born.
JOSHI: What's the difference?
K: To be born is to have a soul.

K thus defines the individual as having a "soul". His belief in humanity can be summarised as: A human is *born* and is thus granted a soul. A replicant is *made* and thus granted nothing. The "soul" here signifies the symbolic—and perhaps for K, even *metaphysical*—difference between human and replicant as being of woman born. Joshi herself reaffirms this distinction as she sends him out on his mission with the rather hollow words of encouragement: "Hey. You're getting on fine without one ... A soul". This belief in the difference between replicant and human is fundamentally what will lead K to question his own certain self-knowledge.

He clings to a childhood memory—of bullies and a carved wooden horse—that he *knows* must be artificial because he had no childhood to be remembered. However, his self-knowledge shifts as he comes to believe a different version of this history. In fact, K comes to believe that this isn't a simple implanted memory, designed—as its creator, Ana, explains—in order to make replicants behave in more predictable ways by giving them mnemic frameworks for processing and understanding emotion. Rather, K comes to believe, following his encounter with Ana, that it is a genuine memory hinting—not unlike with Deckard, to whom it would ultimately connect him—to a repressed truth, expressed once again symptomatically in the image world of the protagonist's psychic reality. This time, however, the situation is reversed: where Deckard's unicorn "dream" disclosed the unconscious knowledge that he was in fact a replicant, here K's horse "memory" seems to disclose the opposite, that he is "human" (or at least is in possession of a *soul*). The reason for K's shifting belief is Ana's ambiguous answer to his question as to whether the horse memory was artificial or in fact remembered. She tells him: "No one invented that. It was a real moment (…) Someone lived this. This happened". To K, this confirms his creeping suspicion that he is *different*, that his self-knowledge is flawed and that there really *is* some part of himself inaccessible to himself. In short, Ana's response convinces him that—despite Wallace's design—the unconscious *is* in play for K: that he is a Lacanian subject (*qua* Thing that thinks) after all.

K has been primed for this belief by his companion, Joi. There are no doubt extensive Lacanian studies to be done of this digital girlfriend as Lacanian Woman (*La femme*), the next stage in the evolution of the masculine-technicist libidinal economy that brought us Samantha in Spike Jonze's *Her* (2013). The latter I have explored elsewhere—and I do not intend to re-tread that argument here—but in short, we can observe that, like Johansson's digital assistant, Joi is another fantasy image: a digital Echo, precision engineered to reflect the desires of the masculine ego back upon itself in a compliant and sexually alluring way.[4] Joi is a virtual cold reader, programmed to pick up on the subtle signs of her "Joe's" personality and present herself accordingly. It should therefore come as no surprise when we recall a prior scene in which K is scanning the genetic archives in search of a missing "orphan": Rachael's seemingly abandoned child, whom K is now beginning to believe is he. She stands at his shoulder and coos in his ear: "I always knew you were special. Maybe this is how. A child of woman born. Pushed into the world. Wanted. Loved". And then, when K visits the orphanage and recovers the wooden horse—concealed two decades before in the ash of the furnace—she declares to him: "I always told you. You're special (…) A real boy now". Like any good Woman, she tells him precisely what he wants to hear, what he wants to believe. And so, despite his certain knowledge to the contrary, K now fully *believes* that he is a "real boy". In fact, I'd even be tempted to claim that he now *believes in the unconscious* (as a sort of inaccessible, mythical origin that would explain his being).

This wavering between certain belief and equally certain knowledge to the contrary is of course what puts K in the realm of the classic *fetishist*: the one who, in Freud's estimation, maintains belief in the maternal phallus even in the traumatic knowledge of its absence. This curious vignette has never made much sense in its own terms but what it *does* usefully point to is the capacity for doubling and splitting belief in perversion, as elaborated of course in Octave Mannoni's famous phrase, "*Je sais bien, mais quand même*...". K *knows very well* that he is a replicant, but at the same time he still *believes* that he is special: that he was born with a soul. The narrative of *2049* is thus driven by K's perverse fantasy of the unconscious itself, grounded in a misinterpretation of the mise-en-scène of memory: it is the trace's author, Ana, who lived this experience, not K. She is the one who was born, wanted, pushed into the world. He is just another Joe. Like Deckard before him, K is haunted by an equine image and, as we discover, Deckard is its ultimate source: having carved the wooden toy for a child (Ana) that he was compelled to abandon in order to secure its future. And like Deckard, this equine image leads to a profound realisation about his nature: it reveals once more that—despite his belief—K is nothing but a replicant.

When Freysa reveals the truth of the horse memory to K, she recognises the shattering pieces of his splintered fantasy: "You imagine it was you? Oh. You did, you did. We all wish it was us. That's why we believe". Where Roy aligned belief to human limitation ("I've seen things you wouldn't believe"), addressing Deckard finally *as if* he were human and insisting on that which he would be incapable of conceiving, Freysa connects belief to a desire for human capacity: to desire itself. But unlike the replicants of *Blade Runner* (Deckard, Rachael)—who initially *believed* themselves to be human and then underwent a degree of subjective destitution through the revelation of their *replicant* identity, reducing them, as Žižek stated, to the pure form of the Lacanian subject (see 1993: 10)—*2049*'s K is simply returned to his initial state of being: not "human" (i.e. born) but just another replicant dreaming of a soul.

He shifts from a perverse position (of the replicant who believes he is human) back into the psychotic structure of the replicant who *knows* he is a replicant.

Villeneuve's film thus conjures similar problematics to its predecessors—drawing on Dick's original inquiry into what it means to be human and grounding this visually (like Scott) in the psychic image world of dream and memory—but finally *2049* shifts the emphasis *away* from this new protagonist in order to rehumanise the original figure, Deckard, in a sense through the position of the Father. K is stripped of his (nascent) humanity, and ultimately his life, as contact with that inaccessible kernel of the unconscious is denied to him once more. In the "Blade Runner" universe, as in *Filmosophy*, then, the unconscious is a *question.* The replicant, as it differs across *Blade Runner* (Deckard) and *Blade Runner 2049* (K), offers two models for conceiving of both the subject's relation to the unconscious and of film-thinking. The replicant is a kind of artificial mind: in *Blade Runner* it is artificial thinking *with* the unconscious (i.e. it does not know it is a replicant); while in *2049*, it is artificial thinking *without* the unconscious (i.e. it knows that it is a replicant, there is no repression). Similarly, filmmaking in the light of *Filmosophy* could even be considered as akin to creating a (fil)mind. But where Frampton would reject the analogy between the *replication* of human thinking and the filmind (the latter not being a *mirror* of human thinking), his conception of filmind (as artificially created organic thought) nonetheless tacks closely to Niander Wallace's conception of the replicant mind *without unconscious* (rather than Tyrell's thinking Things, for example), in that Frampton continually rejects, disavows or mischaracterises the place of the unconscious in film, film theory and his own film-thinking.

In point of fact, Frampton explicitly presents the filmind as thinking without the *subconscious*; however, as the famous joke from Ernst Lubtisch's *Ninotchka* (1939)—"I am sorry sir, but we are out of cream. Could it be [coffee] without milk?", first highlighted by Alenka Zupančič (2005: 173)—insists, what is negated *matters* and has a bearing on the identity of that which is affirmed. In this context, I'd suggest that Frampton's filmind—just like Officer KD6-3.7—is presented as a thinking-without-unconscious (and moreover, that thinking-without-unconscious does not simply equate to conscious thought, as Frampton seems to argue, but to an impoverished understanding of the possibilities of thought itself). However, the traumatically repressed truth in *Filmosophy* (just as it is in *Blade Runner*) *is* the truth of the unconscious: for Deckard it is the truth in the meaning of his daydream of the unicorn (i.e. that he is a replicant); for K it is the *absence* of truth in the horse memory (i.e. that he is still a replicant); while for Frampton, it is the inescapable proximity of his supposedly novel means of thinking film *to* the unconscious (i.e. that filmosophy is, in a sense, an unconscious *replication* of psychoanalytic thought). Film-thinking-*with*-the-unconscious, then, is the model that I wish finally to propose: and, I suggest, it allows us to approach *2049* at a more profound level than Frampton's articulation of the filmosophical filmind would initially permit.

Rethinking *Blade Runner*, 2049

Thinking *Blade Runner 2049* with-the-unconscious, then, should immediately take us towards the question of *form*. Psychoanalysis is, fundamentally, a formalist practice. What is at stake in *The Interpretation of Dreams*, as I have noted, is the relation between the content of the dream (both manifest and latent) and the "disguised wish"—or *desire*—that cuts across the content, in the process of the dreamwork, and distorts its form in the very production of the dream. Moreover, the central aim of Lacanian psychoanalysis is, as Lacan explains in Seminar XI, "the elaboration of the notion of the subject" (1978: 77). It is an approach to the question of the subject in terms of its *form*, which is to say at the level of the *cogito*. Lacan shows the Freudian unconscious to be a subversion of the Cartesian subject, while in fact remaining Cartesian in its origin. It turns the *cogito* inside-out because it is the unconscious that "thinks" on *ein anderer Schauplatz*. Lacan reformulates Descartes thus: "I am thinking where I am not, therefore I am where I am not thinking", in the extimate big Other (2007: 430). The "I" of "I am thinking" is therefore split, between "the level of the enunciation (*énonciation*) [and] the level of the statement (*énoncé*)" (Lacan 1978: 139). The subject of the enunciation is the subject of the unconscious: not present in the statement but presupposed by it, whereas the subject of the statement is the self-conscious "I" of the ego. The subject of the unconscious is thus the subject at the level of *pure form*: the "Thing that thinks" of the cogito, to which I alluded above and that Žižek aligns with the basic subjective level of the replicant in *Blade Runner*.

By the same token, the great benefit of Frampton's project in *Filmosophy* is certainly the (re)turn towards form that he insists upon in the philosophical consideration of film (i.e. *film thinks in form*). Where such a theoretically-oriented mode of inquiry focuses on *concepts* there is sometimes a tendency to discern them at the level of narrative, theme and characterisation (the arguably more "literary" aspects of film, at which *ideas* might seem more readily available). However, rather than simply rejecting so-called "content analysis" in film, Frampton notes: "Form has still been seen as separate, usually brought in only when its actions confirm an interpretation of the film's story. This last point is important, as the route to interpretation should always be via the whole film, not biasing form or content" (2006: 169). In parallel to this, McGowan also insists on an approach to cinema at a formal level. He suggests that film is created by a filmmaker in order to appeal to the desire of the spectator, and works this through in relation to the psychoanalytic understanding of the dreamwork (as I have already suggested) as a formal construction giving expression to *desire*: the filmmaker working like the unconscious itself, creating an experience for the subject in sound and image.

The connection between dream and film at the level of form is of course crucial (and has been recognised by the earliest film theorists from Maxim Gorky's "Kingdom of Shadows" onwards). However, what McGowan offers is effectively a theory of *spectatorship*—of the viewer/unconscious as it encounters the film—while what I am seeking here is a theory of the *film form itself* grounded in the same principles. It is not so much that the form of the film holds the secret of either the desire of the filmmaker or the spectator but that it holds the secret of the film(ind) itself, where the latter is understood as a Lacanian reframing of Frampton's film-thinking: as a kind of thinking thing analysable along the same lines as the Thing that thinks. As McGowan insists, "psychoanalytic interpretation involves examining the form of the film as it relates to the unconscious in some way" (2015: 11). And this is undoubtedly true for the spectator but I suggest we could take such a formulation even further: as I have already begun to suggest, we should approach film *as we approach the unconscious*, not only to discern the desires to which it might speak in the spectator but also—crucially—to consider how we might conceive of the film as (being akin to) the unconscious, itself, speaking.

I am not, to reiterate, relying on the idea of the unconscious of the artist manifesting through the art—filmmakers may well reveal certain aspects of themselves in their work, of which they are themselves unaware—but, rather, I am claiming that we should conceive of film itself—along the lines of Frampton *avec* Lacan—as a kind of "thinking being". This would not be the "rational agent" of the modern cogito but the properly Lacanian cogito—the subject of the unconscious—that expresses itself through sounds and images (much like the dreamwork) but nonetheless remains opaque to itself (and others) except where careful attention is paid to the particular affects and signifiers presented. As I proposed above, an encounter with film could be conceptualised as being like an encounter with the *unthinking* of the unconscious: it obeys its own logic(s) and signifies it its own ways. The interpretation of films, then, really is like the interpretation of dreams in that each formation much be taken on its own terms, requiring a skilful analyst to determine what is at stake and—crucially, this is what I would add to my original proposition—to understand the ways in which the film speaks *in excess of itself*, as the *ça parle* of a filmind-unconscious.

Approaching *2049*, then, directly at the level of form, there is—I'd argue—an overwhelming sense in which it presents as a sort of *replicant film*: a (slightly vacant) simulacrum of an earlier model. Indeed, *2049* positions itself overtly as a self-conscious "replication" of *Blade Runner*. Apart from being a direct sequel to Scott's film—which of course entails any number of continuities in terms of style and narrative: the replicants and their implanted memories, bounty hunting blade runners, artificial animals, and holographic adverts of post-apocalyptic Los Angeles and the off-world colonies, etc., Eldon Tyrell resurrected as Niander Wallace, and Deckard, Rachael and Gaff all returning in various forms—Villeneuve's contribution is littered with explicit references to the sounds and images of *Blade Runner*: K's use of the "pilot fish" drone to scan and enhance images of Las Vegas evokes Deckard's ESPER machine; the surf-sprayed nocturnal showdown between K and Luv restages Roy and Deckards's own rain-soaked conflict, as in fact does the *mano-a-mano* combat between K and Sapper Morton that opens the film; the bees that cover K's hand in Las Vegas recall Deckard's question during Rachael's Voight-Kampff test regarding a wasp crawling on her arm (but where she exclaims that she would kill it, he allows them to settle harmlessly around his fingers); both films feature extreme close-ups on the eye in their opening sequences; and Hans Zimmer's score frequently recalls Vangelis' electronic instrumentation, explicitly quoting the melody of "Tears In Rain" as K lays down to die (like Roy) at the end of *2049*.

And while, as I have already noted, each protagonist is haunted by an equine image (Deckard's unicorn, K's wooden horse) that seems to point to the fundamental question of identity, these other instances add up to a collection of surface images rather than anything more profound. This is perhaps best exemplified by Gaff, who continues in *2049* to populate his paper menagerie with origami specimens but where, as I have noted, he leaves Deckard a symptomatic *unicorn*, for K he constructs a knowingly metatextual *sheep* as a nod to Dick's novel. This would of course be very much in keeping with the terms of Scott's *Blade Runner* as postmodern text par excellence, which presents a Jamesonian *pastiche* of '40s noir stylings: the future-retro of 2019 offering *empty parody* of historical moods and a depthless play of images. Indeed, charting the fully seven different versions of Scott's film, Flisfeder suggests that "From *Electric Sheep?* To *The Final Cut*, *Blade Runner* is inherently a product of postmodernity, a constant simulacrum of itself. It is impossible that any one version is more authentic than the others; it is also difficult to say which one is truly the 'original'" (2017: 97). Now *2049* offers a further iteration of this logic: yet another copy of a copy of a copy without true origin. Ultimately, *2049* seems to eschew the existential dilemmas that drove its predecessors: constantly skirting the question of Deckard's status, the question is deflected instead to his dog. K wonders whether it is real or artificial but Deckard's only response is, "I don't know, ask *him*", rejecting the question of origins once and for all in favour of the simulacra.

This replicant logic in *2049* finds its purest expression in the most overt call back to *Blade Runner*: the (re)appearance of Rachael in

Niander Wallace's chamber. The main innovations of Villeneuve's work are produced by "retconning" the 1982 film (inserting certain elements into *Blade Runner*'s narrative in order to provide retroactive continuity with the present story). Wallace thus rewrites the encounter between Rachael and Deckard in Tyrell's office as an event engineered in order to cause them to fall in love and produce a child, recasting her as the biblical Rachel who was infertile until God granted her a son. First introduced in *2049* as an anonymous box of bones buried beneath Sapper Morton's tree, and then as a fragment of a voice recording in the Tyrell/Wallace archives, Rachael now returns to life as Deckard's promised reward if he will betray their child. A distinctive silhouette scales the steps of Wallace's chamber and glides towards Deckard: at first the effect is so convincing that it seems unclear whether this is in fact a flashback to *Blade Runner* or a new staging for *2049*. The dress, the hair, the lipstick, Rachael looks just like that day when we first met her: as if she has emerged directly from Deckard's memory (or fantasy). Like *2049* itself, she is a beautiful reproduction of the 2019 model of Scott's film. But the effect does not quite *work*. There is something slightly *off* about her: the shape of her mouth, the play of light on her skin, her very presence in the chamber, none of it wholly convinces. The digital overlay of Sean Young's face from 1982 onto an actor in 2017 sits as awkwardly as the Rachael of 2019 does now in the new world of *2049*.

Strikingly, this *failure* of the VFX here resonates powerfully with her diegetic significance: Rachael's presence is not quite right for Deckard—"Her eyes were green", he notes—and it is not quite right for the spectator either: more than just eye colour, she does not traverse the uncanny valley from simulacrum to human. Freud famously suggested that the return of the dead was one of the many avatars of *das Unheimliche* and here there is, I might say, a *double uncanny* at play: Deckard's repulsion in the face of Wallace's "gift" is matched by the spectator's disquiet in witnessing Sean Young's digital revenant. The slopes of that valley here might be said to work in the film's favour, resonating shock and unease from the diegesis outwards. In the existing terms of film analysis, we might observe that Rachael thus becomes what Kristen Whissel calls an *emblem*: an instance in which VFX directly embody the themes and narrative concerns of the film (see 2014), the double uncanny of

Rachael's presence triggering a similar response in the spectator as it does in Deckard. However, there is also an extra dimension at play here, which takes this effect *beyond* the diegesis and towards a commentary on *2049* overall. This moment signifies far *in excess* of the film itself, pointing to the fundamental logic that determines its organisation: and this is where it is vital to discern the film-thinking-with-the-unconscious beyond *Filmosophy* that I am proposing.

Frampton rightly criticises approaches to film analysis that see elements of style as a deviation from a formal norm, rather than as "thinkings in their own right" (2006: 108). He notes that "cognitivists always see style in relation to a base normality (while for filmosophy it is exactly this 'beyond' that is of interest)" (2006: 108). However, I would argue that a "beyond" or *excess* in film is precisely what filmosophy fails to conceptualise. As I have already noted, Frampton repeatedly frames film-thinking in explicitly phenomenological terms, in which the film image is *intended* by the filmind; however, there is a significant tendency in the language of *Filmosophy* towards a more everyday understanding of "intentionality". For example, in the showpiece chapter on "film-thinking", Frampton asserts apropos of the close-up that "The filmind *wants us* to know how important the preceding moment is for the character involved, or *wants to make us* ask ourselves what it means for another, seemingly peripheral character" (2006: 128; emphasis added). In effect, every part actively contributes to a unified, meaningful totality of the film. In response, the psychoanalytic question—posed by film-thinking-with-the-unconscious—must be: What about (film-)thinking that is *in excess of itself*, giving rise to meanings and affects perhaps even in spite of itself? Indeed, where Frampton insists—in a filmosophical restating of the classic notion of the syuzhet or story—that the filmind "[has] the whole film in its knowledge" (2006: 137) again I wonder: What is *not* in the filmind's "knowledge" (as it is formulated in *Filmosophy*)? Of what order is the filmind's *non-knowledge* (i.e. the unconscious)?

In this light, I would contend that filmosophy would not be able to account for the effect of Rachael's presence in *2049*. In short, I am claiming that Frampton's model of the filmind leaves no place whatsoever for such excess. If we can consider Rachael as the *emblem* of

2049's replicant filmind, then she is a visual manifestation of the depthless, iterative logic of the film itself. Rather than simply presenting a synthesis of form and content in *2049*, however, Rachael's appearance signifies *extra-textually*: pointing towards the general problem of Villeneuve's film. The failure of the VFX here—in the sense that, while within the diegesis Rachael is presented as a flesh-and-blood synthetic human like any other replicant, perceptually she lacks the photorealistic presence onscreen offered by live action counterparts such as Wallace and Deckard—matches a certain failure in the overall experience of *2049*. As critics have observed, the film feels almost as empty as Sean Young's digital facsimile: Jake Coyle calls *2049* "resolutely gorgeous" but "broken" (2017), and K Austin Collins deems it "a Stylish but Hollow Spectacle" (2017), for example. Most notably Richard Lawson in *Vanity Fair* comments on the film's "Jaw-Dropping Style but Too Little Substance", observing that Villeneuve "tends to create gorgeous, but rather empty vessels"; and where he notes that "Denis Villeneuve's sci-fi sequel looks amazing, yet left us cold", he is not explicitly referring to Rachael but could just as easily be relating Deckard's encounter with her simulacrum (2017). Like their creation Niander Wallace, Villeneuve and Roger Deakins have a flair for style and precision-engineered spectacle but, as Collins again notes, "it feels a bit empty" (2017). Such determinations would be beyond the scope of *Filmosophy*, where this meta-textual aesthetic compromise would have to be conceived of as *part* of the film's intended totality, rather than the symptomal point within the form that signifies its greater (*unintended*) truth. The *2049* filmind may have given the ersatz Rachael an uncanny, not-quite-human-enough CGI patina but it did not *decide* to index this to the overall compromises of the work: the effect is *in excess* of whatever it might have "intended". The part really does contain the truth of the whole here but not, perhaps, as Frampton might have supposed.

The unconscious, then, figures first of all in *Filmosophy* as a blind spot—its position cannot be countenanced in the elaboration of Frampton's project—and yet, I argue, much of what he claims for filmosophy can equally be claimed for psychoanalysis (the unconscious as trans-subjective no-place, for example). But moreover, film analysis also requires a place for the unconscious, for what *exceeds* the terms of the cinematic experience. These two claims thus have different implications for our understanding of filmosophy. The first—what is said about the filmind can *also* be said about the unconscious—would fundamentally leave Frampton's project unchanged: there might be a slight psychoanalytic corrective to the way that *Filmosophy* is situated in relation to the field but the two discourses could exist in harmonious parallel. The second—that the filmind requires an understanding of film-thinking-with-the-unconscious—*does* necessitate a re-evaluation of foundational terms and concepts: *Filmosophy* cannot emerge unscathed from this encounter with Lacan and *Blade Runner 2049*. As the resurrection of Rachael demonstrates, where the film operates in excess of itself what is required is an appreciation of the intentionality of the *unconscious* rather than a notion of film-thinking as intentional totality. Filminds, then, *really do dream of celluloid sheep*.

Conclusion: The Electric Dream

Frampton's conception of film-thinking—like Niander Wallace's conception of the compliant replicant in *2049*—leaves no place for a (filmind) unconscious, for any locus outside of the conscious intention of the thinking being: "every frame is intended" (2006: 84). *Filmosophy* constitutes the film as a meaningful totality, a mind that has total knowledge of its content, whereas Lacanian psychoanalysis would emphasise the very *form* of this knowing in the empty place of the cogito. I have argued that, in *2049*, an example of this effect of excess to the intentionality of the filmind can be found in Rachael's resurrection. The "failure" of the film form in this instance to transcend its digital limitations certainly evokes the thematics of Villeneuve's film and re-emphasises Deckard's shock and disgust in the face of this pale imitation of his lover. And if that were the full extent of the effect, this would no doubt remain congruent with Frampton's vision of the absolute synthesis of form and content in the encounter with a filmind. However, as I have argued, the impact extends *beyond* the text itself and can be found reverberating in the popular and critical response to the film: in particular, it seems as though viewers of Scott's iconic—if flawed—postmodern classic are similarly dismayed at the visage of *Blade Runner 2049*: its too-perfect surfaces concealing only the lack of depth beneath. Where Dick's originary enquiries into the "electric dream" were transcribed into *Blade Runner*'s meditations on mortality and identity, *2049* reduces the question to notions of natality and paternity. And where, as I have explored, previous iterations of the replicant put the properly Lacanian subject in play (as the pure form of the cogito *qua* Thing that thinks) both *2049* and *Filmosophy* seem to rely on specifically limited forms of mind, modelled on the notionally self-transparent cogito of modern philosophy. It has been the gaps, oversights and lacunae in this latter understanding of "mind" that I have attempted to highlight here and, in the final analysis, what *Filmosophy* and *Blade Runner 2049* alike insist upon is a return to the dream, the memory, the fantasy: to the truth of the unconscious as it is found in Lacanian psychoanalysis.

Notes

1. See *Out of the Past: Lacan and Film Noir* (Tyrer 2016: 6–10).
2. See "*Under Her Skin*: On Woman without body and body without Woman" (Tyrer 2019).
3. This is one way of understanding Žižek's famous joke about the man "cured" of his belief that he is a grain of corn: "I know I'm not but has anyone told the chicken?!". The truth must be subjectively assumed in the unconscious ("by the chicken") in order to be operative. See Žižek (2006: 351).
4. See Tyrer (2019).

Bibliography

Collins, K. A. (2017) 'Blade Runner 2049' Is a Stylish but Hollow Spectacle. *The Ringer*. Available from: https://www.theringer.com/movies/2017/10/5/16429424/blade-runner-2049-film-review-ryan-gosling-harrison-ford-denis-villeneuve. [Accessed 22 February 2018].

Coyle, J. (2017) Review: Gorgeous '2049' breaks the 'Blade Runner' spell. *AP News*. Available from: https://apnews.com/b377bb2396e647a6bf373de2aa8f4535. [Accessed 22 February 2018].
Dolar, M. (1998) Cogito as the Subject of the Conscious. In S. Žižek (ed.) *Cogito and the Unconscious*. Durham: Duke University Press.
Fisher, M. (2016) *The Weird and the Eerie*. Watkins Media: Kindle Edition.
Flisfeder, M. (2017) *Postmodern Theory and Blade Runner*. New York: Bloomsbury.
Frampton, D. (2006) *Filmosophy*. London: Wallflower Press.
Gallop, J. (1987) *Reading Lacan*. Ithaca: Cornell University Press.
Lacan, J. (1977) *Écrits: A Selection*. Trans. by Alan Sherdian. London: Tavistock.
Lacan, J. (1978) *The Four Fundamental Concepts of Psychoanalysis*. Trans. by A. Sheridan. New York: Norton.
Lacan, J. (1993) *The Psychoses*. Trans. by R. Grigg. New York: Norton.
Lacan, J. (1997) *The Ethics of Psychoanalysis*. Trans. by D. Porter. New York: Norton.
Lacan, J. (2007) *Écrits*. Trans. by B. Fink. New York: Norton.
Lawson, R. (2017) Blade Runner 2049 Review: Jaw-Dropping Style but Too Little Substance. *Vanity Fair*. Available from: https://www.vanityfair.com/hollywood/2017/10/blade-runner-2049-review. [Accessed 22 February 2018].
McGowan, T. (2015) *Psychoanalytic Film Theory and The Rules of the Game*. New York: Bloomsbury.
Miller, J.A. (1994) Extimité. In Mark Bracher et al. (eds.) *Lacanian Theory of Discourse*. New York: New York University Press.
Neill, C. (2018) Do Electric Sheep Dream of Androids? On the Place of Fantasy in the Consideration of the Nonhuman. In G. Basu Thakur and J.M. Dickstein (eds.) *Lacan and the Nonhuman*. London: Palgrave Macmillan.
Tyrer, B. (2016) *Out of the Past: Lacan and Film Noir*. London: Palgrave Macmillan.
Tyrer, B. (2019) Under Her Skin: On Woman Without Body and Body Without Woman. In A. Piotrowska and B. Tyrer (eds.) *Femininity and Psychoanalysis: Cinema, Culture, Theory*. London: Routledge.
Whissel, K. (2014) *Spectacular Digital Effects: CGI and Contemporary Cinema*. Durham: Duke University Press.
Žižek, S. (1992) *Looking Awry*. Cambridge: MIT Press.

Žižek, S. (1993) *Tarrying with the Negative*. Durham: Duke University Press.
Žižek, S. (2006) *The Parallax View*. Cambridge: MIT Press.
Zupančič, A. (2005) Reversals of Nothing: The Case of the Sneezing Corpse. *Filozofski vestnik*, 2, 173–186.

3

Blade Runner 2049: A View of Post-Human Capitalism

Slavoj Žižek

How are capitalism and the prospect of post-humanity related? Usually it is posited that capitalism is (more) historical and our humanity, inclusive of sexual difference, more basic, even ahistorical; however, what we are witnessing today is nothing less than an attempt to integrate the passage to post-humanity into capitalism. This is what the efforts of new billionaire gurus like Elon Musk are about; their prediction that capitalism "as we know it" is coming to an end refers to "human" capitalism, and the passage they talk about is the passage from "human" to post-human capitalism. *Blade Runner 2049* deals with this topic—here is the storyline (shamelessly borrowed from Wikipedia).[1]

In 2049, replicants (bioengineered humans) have been integrated into society as servants and slaves. K, a newer replicant model who is created to obey, works as a "blade runner" for the LAPD, hunting down and "retiring" rogue older model replicants. His home life is spent with his holographic girlfriend Joi, an artificial intelligence product of the

S. Žižek (✉)
Department of Philosophy, University of Ljubljana, Ljubljana, Slovenia

C. Neill (ed.), *Lacanian Perspectives on Blade Runner 2049*,
The Palgrave Lacan Series,
https://doi.org/10.1007/978-3-030-56754-5_3

Wallace Corporation. K's investigation into a growing replicant freedom movement leads him to a farm, where he retires rogue replicant Sapper Morton and finds a buried box. Forensic analysis reveals the box contains the remains of a female replicant who died as the result of complications from an emergency caesarean section. K finds this unsettling as pregnancy in replicants was originally thought to be impossible.

K is ordered by his superior, Lieutenant Joshi, to destroy all evidence related to the case and to retire the child. Joshi believes the knowledge that replicants are able to reproduce to be dangerous and that it could lead to war, since it blurs the clear line of separation between replicants and humans. K, disturbed by his orders to kill a born individual, visits the headquarters of the Wallace Corporation and meets its founder Niander Wallace who identifies the body as Rachael, an experimental replicant. In the process, he learns of her romantic ties with former veteran blade runner Rick Deckard. Believing that reproduction in replicants can bolster his production and expand his off-world operations, but lacking the technology to give them this ability himself, Wallace sends his replicant enforcer Luv to steal Rachael's remains from the LAPD headquarters and to follow K to find Rachael's child.

Returning to Morton's farm, K finds a hidden date that matches a childhood memory about hiding a toy horse, which he later finds at an orphanage, suggesting that his memories—which he thought were implants—are real; Joi insists this is evidence that K is in fact a real person. While searching birth records for that year, he discovers that twins were born on that day with identical DNA except for the sex chromosome; only the boy is listed as alive. K seeks out Dr. Ana Stelline, a memory designer for the Wallace Corporation, who informs him that it is illegal to program replicants with humans' real memories, leading K to believe he might be Rachael's son. After failing a test of his replicant behavior, K is suspended by Joshi, but Joshi gives him 48 h to disappear. After transferring Joi to a mobile emitter, despite knowing that if it is damaged she will be erased, K has the toy horse analyzed and finds traces of radiation that lead him to the ruins of Las Vegas, where he finds Deckard. Deckard reveals that he scrambled the birth records to cover his tracks and was forced to leave a pregnant Rachael with the replicant freedom movement to protect her.

Luv and her men murder Joshi, track K's location and arrive to kidnap Deckard. They leave a badly injured K for dead and destroy Joi's emitter. He is later rescued by the replicant freedom movement who were also tracking him. He is told by their leader, Freysa, that he is wrong to think he has a unique role to fulfil in the movement, and that Rachael's child is actually a girl. K deduces that Stelline is Deckard's daughter, as she is the only one capable of creating the memory and implanting it into him. Freysa urges K to prevent Wallace from uncovering the secrets of replicant reproduction by any means necessary, including killing Deckard.

In Los Angeles, Deckard is brought before Wallace, who suggests Rachael's feelings for him were engineered by Tyrell to test the possibility of a replicant becoming pregnant. When Deckard refuses to cooperate, Wallace has Luv escort him to off-world outposts to be tortured for information, but K intercepts them, killing Luv and staging Deckard's death to protect him from both Wallace and the replicants. He leads Deckard to Stelline's office and laments that his best memories belong to her. Deckard cautiously enters the office and approaches Stelline, while K succumbs to his wounds…

So why is the fact that two replicants (Deckard and Rachael) formed a sexual couple and created a human being in a human way, experienced as such a traumatic event, celebrated by some as a miracle and castigated by others as a threat? Is it about reproduction or about sex, i.e., about sexuality in its specific human form? The movie focuses exclusively on reproduction, again neglecting the big question: can sexuality, deprived of its reproductive function, survive into the post-human era? The image of sexuality remains the standard one: sexual act is shown from the male perspective, so that the flesh-and-blood android woman is reduced to the material support of the hologram fantasy-woman Joi, created to serve the man: "she must overlap with an actual person's body, so she is constantly slipping between the two identities, showing that the woman is the real divided subject, and the flesh and blood other just serves as a vehicle for the fantasy" (McGowan). The sex scene in the film is thus almost too directly "Lacanian" (in line with films like *Her*), ignoring authentic hetero-sexuality (in which the partner is not just a support for me to enact my fantasies but a real Other).[2] The movie also fails to explore the (potentially antagonistic) difference among androids themselves, between the "real flesh" androids and the android whose body is just a 3-D hologram projection: how does, in the sex scene, the flesh-and-blood android woman relate to being reduced to the material support of the male fantasy? Why doesn't she resist and sabotage it?

The movie provides a whole panoply of modes of exploitation, inclusive of a half-illegal entrepreneur using child labor (hundreds of human orphans) to scavenge old digital machinery. From a traditional Marxist standpoint, strange questions arise here: if fabricated androids work, is exploitation still operative here? Does their work produce value which is in excess of their own value as commodities so that it can be appropriated by their owners as surplus-value? One should note that the idea of enhancing human capacities in order to create post-human perfect workers or soldiers has a long history in the XXth century. In the late 1920s, none other than Stalin for some time financially supported the "human-ape" project proposed by the biologist Ilya Ivanov (a follower of Bogdanov, the target of Lenin's critique in *Materialism and Empirio-Criticism*). The idea was that by way of coupling humans and orangutans, one will create a perfect worker and soldier, impervious to pain, tiredness and bad food. In his spontaneous racism and sexism, Ivanov, of course, tried to couple male humans and female apes, and the humans he used were black males from Congo since they were supposed to be genetically closer to apes—the Soviet state financed an expensive expedition to Congo. When his experiments failed, Ivanov was liquidated. Furthermore, Nazis also regularly used drugs to enhance the fitness of their elite soldiers. The US army is now experimenting with genetic changes and drugs to make soldiers super-resilient (they already have pilots ready to fly and fight 72 hours, etc.). In the domain of fiction, one should include zombies on this list. Horror movies register class difference in the guise of the difference between vampires and zombies: vampires are well mannered, exquisite, aristocratic, they live among normal people, while zombies are clumsy, inert, dirty, and attack from the outside, like a primitive revolt of the excluded. The equation between zombies and the working class was directly made in *White Zombie* (1932, Victor Halperin), the pre-hays-Code first full-length zombie film. There are no vampires in this film—but, significantly, the main villain who controls zombies is played by Bela Lugosi who had become famous a year earlier playing Dracula. *White Zombie* takes place on a plantation in Haiti, the site of the most famous slave revolt. Lugosi receives another plantation owner and shows him his sugar factory where workers are zombies who, as Lugosi is quick to explain, don't complain about long working hours, demand no trade unions, never strike, but just go on and on working… such a film was possible only before the imposition of the Hays Code.

In yet another nice reversal of the standard formula in which the hero, living as (and thinking he is) just an ordinary guy, discovers he is an exceptional figure with a special mission. Here K thinks he is the special figure everybody is looking for (the child of Deckard and Rachael), but gradually realizes that (as with many other replicants) he is just an ordinary replicant obsessed with an illusion of greatness, so he ends up sacrificing himself for Stelline, the true exceptional figure everyone is looking for. The enigmatic figure of Stelline is crucial here: she is the "real" (human) daughter of Deckard and Rachael (the result of their copulation), which means a human daughter of replicants, inverting the process of man-made replicants. Living in her isolated world (unable to survive in open space filled with real plants and animal life), contained to utter sterility (white dress in an empty room with white walls), her contact with life limited to the virtual universe generated by digital machines, she is ideally positioned as a creator of dreams (she works as an independent contractor, programming false memories to be implanted into replicants). As such, Stelline exemplifies the absence (or, rather, impossibility) of the sexual relationship, which she supplants with a rich fantasmatic tapestry. No wonder that the couple which is created at the film's end is not the standard sexual couple but the asexual couple of father and daughter. This is why the final shots of the film are so familiar and weird at the same time: K sacrifices himself in a Christ-like gesture on snow to create the couple… of father and daughter.

Is there a redemptive power in this reunion? Or should we read the fascination with this reunion against the background of the film's symptomatic silence about social antagonisms among humans in the society it depicts: where do human "lower classes" stand? However, the movie does render nicely the antagonism that cuts across the ruling elite itself in our global capitalism: the antagonism between State and its apparatuses (personified in Joshi) and big corporations (personified in Wallace) pursuing progress to its self-destructive end. "While the state political-legal position of the LAPD is one of potential conflict, Wallace sees only the revolutionary productive potentials of self-reproducing replicants, which he hopes could give him a leg up in his business. His perspective is one of the market; and it is worth looking at these contradictory perspectives of Joshi and Wallace, for it is indicative of the contradictions that do exist between the political and the economic; or, put differently, it oddly indicates the intersection of the class state mechanism and the tensions in the economic mode of production (Flisfeder).

Although Wallace is a real human, he already acts as inhuman, an android blinded by excessive desire, while Joshi stands for apartheid, for the strict separation of humans and replicants—her standpoint is that, if this separation is not upheld, there is war and disintegration: "If a child is born from a replicant mother (or parents), does he remain a replicant? If he has produced his own memories, is he still a replicant? What is now the dividing line between humans and replicants if the latter can self-reproduce? What marks our humanity?" (Ibid.).

So should we not, with regard to *Blade Runner 49*, supplement the famous description from *The Communist Manifesto*, adding that also sexual "one-sidedness and narrow-mindedness become more and more impossible," that also in the domain of sexual practices, "all that is solid melts into air, all that is holy is profaned" (Marx & Engels), so that capitalism tends to replace the standard normative heterosexuality with a proliferation of unstable shifting identities and/or orientations? Today's celebration of "minorities" and "marginals" IS the predominant majority position—even alt-rightists who complain about the terror of liberal Political Correctness present themselves as protectors of an endangered minority. Or take those critics of patriarchy who attack it as if it is still a hegemonic position, ignoring what Marx and Engels wrote more than 150 years ago, in the first chapter of *The Communist Manifesto*: "The bourgeoisie, wherever it has got the upper hand, has put an end to all feudal, *patriarchal*, idyllic relations."—is still ignored by those Leftist cultural theorists who focus their critique on patriarchal ideology and practice. Not to mention the prospect of new forms of android (genetically or biochemically manipulated) post-humanity which will shatter the very separation between human and inhuman.

So why does the new generation of replicants not rebel? "Unlike the replicants in the original, the newer replicants never revolt, though it is not clearly explained why, other than they are programmed not to. The film, however, hints at the explanation: the fundamental difference between the new and old replicants involves their relation to their false memories. The older replicants revolted because they believed their memories to be real and thus could experience the alienation of recognizing that they weren't. The new replicants know from the beginning that their memories are faked, so they are never deceived. The point is thus that fetishistic disavowal of ideology renders subjects more enslaved to the ideology than simple ignorance of its functioning" (McGowan). The new generation of replicants is deprived of the illusion of authentic memories, of all substantial content of their being, and thereby reduced to the void of subjectivity, i.e., to the pure proletarian status of *substanzlose Subjektivitaet*. So, does the fact that they don't rebel mean that rebellion has to be sustained by some minimal substantial content threatened by the oppressive power?

K stages a fake accident to make Deckard disappear not only from the sight of state and capital (Wallace) but also from the sight of the replicant rebels (who are led by a woman, Freysa, a name which, of course, echoes freedom, *Freiheit* in German). Although one can justify his decision by the fact that Freysa also wants Deckard dead (so that Wallace will not be able to discover the secret of the replicant reproduction)—both State apparatus (embodied in Joshi) and revolutionaries (embodied in Freysa) want Deckard dead—, K's decision nonetheless gives to the story a conservative-humanist twist: it tries to exempt the domain of family from the key social conflict, presenting both sides as equally brutal. This not-taking sides betrays the falsity of the film: it is all too humanist, in the sense that it all circulates around humans and those who want to be (taken as) humans or those who don't know if they are not humans. (Is the result of biogenetics not that we—"ordinary" humans—effectively are that, humans who don't know they are not humans, i.e., neuronal machines with self-awareness?). The film's implicit humanist message is that of liberal tolerance: we should give androids with human feelings (love, etc.) human rights, threat them like humans, incorporate them into our universe—but will, with their arrival, our universe still be ours, will it remain the same human universe? What is missing is any consideration of the change that the arrival of androids with awareness will mean for the status of humans themselves: we humans will no longer be humans in the usual sense, something new will emerge, so how to define it? Furthermore, with regard to the distinction between androids with "real" body and hologram androids, how far should our recognition extend? Should also hologram replicants with emotions and awareness (like Joi who was created to serve and satisfy K) be recognized as entities which act as humans? We should bear in mind that Joi, ontologically a mere hologram replicant with no actual body of its own, commits in the film the radical act of sacrificing herself for K, an act for which it (or, rather, she) was not programmed.[3]

Avoiding this new world leaves only the option of a nostalgic feeling of threat (the threatened "private" sphere of sexual reproduction), and this falsity is inscribed into the very (visual and narrative) form of the film in which the repressed of its content returns: not in the sense that the form is more progressive, but in the sense that the form serves to obfuscate the progressive anti-capitalist potential of the story. The slow rhythm with aestheticized imagery directly expresses the social stance of not-taking-sides, of passive drifting.

So what would have been an authentic contact between a human and a replicant? Let's take a (perhaps) surprising example: *Wind River* (Taylor Sheridan, 2017), a movie which tells the story of Natalie Hanson, a native American girl found raped and frozen in the middle of Winter on a desolate Wyoming reservation. Cory, a hunter whose girl also disappeared three years ago, and Jane, a young FBI agent, try to unravel the mystery. In the final scene, Cory goes to Hanson's house where he finds a desperate Martin, Natalie's father, sitting outside with a "death face" (a mix of blue and white) painted on his face. Cory asks him how he learnt to do it, to which Martin replies: "I don't. I've just made it up. It's no one left to teach it." He informs Cory that he just wanted to let it all go and die when a phone rang—his (delinquent) son Chip called him, released from prison, asking him to pick him up at the bus station. Martin says he will do it "as soon as I wash this shit off my face": "I should go and get him, eventually. Just seat here for a minute. Get time to seat with me?" Cory says yes, they sit there silently, and a title screen comes up saying that statistics are kept for every group of missing people except native American women. Nobody knows how many are missing. The terse beauty of this ending is slightly damaged only by these final words on the screen (they state the obvious and thus introduce an element of fake objectivity into an extreme existential drama.) The underlying problem is that of a ritual of mourning which enables us to survive an unbearably traumatic loss, and the glimmer of hope provided by the ending is that Martin and Cory will be able to survive through such a minimal ritual (of just sitting silently). We should not dismiss lightly Martin's "as soon as I wash this shit off my face" as grounded in the fact that his death face is not in the old authentic way but has just been improvised by him: it would remain "shit" even if it were to be done authentically. Martin has already irretrievably lost his ancient ethnic substance, he is already a modern subject unable to practice "death face" with full immersion—however, the miracle is that, although he knows and assumes all this, improvising a death face and just sitting there with it works as authentic in its very artificial improvisation—it may be shit, but shit works in its very minimal gesture of withdrawal from life engagement. We should bear in mind here that Cory is a white man living in a reservation, and what Martin asks him to do is not to show solidarity with a grieving native American and participate in a ritual which is meaningless to him—such patronizing respect for a primitive culture is one of the most disgusting versions of racism. The message of Martin's request is that he shares with Cory the distance the latter feels towards the native American ritual: Cory's—white man's—distance is already Martin's, and it is this distance which makes the ritual authentic, not part of some ridiculous "immersing into a native culture." Do we not encounter here yet another example of a twist that can be characterized as a Moebius strip? When we progress from the naïve immersion in a ritual to its utter dismissal as something ridiculous, we all of a sudden find ourselves back in the same ritual, and the fact that we know it is all rubbish in no way diminishes its efficiency.

So can we imagine something homologous taking place between a human and a replicant? A situation in which the two invent (and participate in) a similar empty ritual? A ritual which is in itself totally meaningless—we search in vain for a deeper message hidden in it—since its function is purely tautological (or what Jakobson called phatic)?

When the question "are androids to be treated like humans?" is debated, the focus is usually on the awareness or consciousness: do they have an inner life? (Even if their memories are programmed and implanted, they can still be experienced as authentic.) Perhaps, however, we should change the focus from consciousness or awareness to the unconscious: do they have an unconscious in the precise Freudian sense? The unconscious is not some deeper irrational dimension but what Lacan would have called a virtual "another scene" which accompanies the subject's conscious content. Let's take a perhaps unexpected example. Recall the famous joke from Lubitsch's *Ninotchka*: "'Waiter! A cup of coffee without cream, please!' 'I'm sorry, sir, we have no cream, only milk, so can it be a coffee without milk?'" At the factual level, coffee remains the same coffee, but what we can change is to make the coffee without cream into a coffee without milk—or, more simply even, to add the implied negation and to make the plain coffee into a coffee without milk. The difference between "plain coffee" and "coffee without milk" is purely virtual, there is no difference in the real cup of coffee, and exactly the same goes for the Freudian unconscious: its status is also purely virtual, it is not a "deeper" psychic reality—in short, unconscious is like "milk" in "coffee without milk." And therein resides the catch: can the digital big Other which knows us better than we know ourselves also discern the difference between "plain coffee" and "coffee without milk"? Or is the counterfactual sphere outside the scope of the digital big Other which is constrained to facts in our brain and social environs that we are unaware of? The difference we are dealing with here is the difference between the "unconscious" (neuronal, social…) facts that determine us and the Freudian "unconscious" whose status is purely counterfactual. This domain of counterfactuals can only be operative if subjectivity is here: in order to register the difference between "plain coffee" and "coffee without milk," a subject has to be operative. And, back to *Blade Runner 49*, can replicants register this difference?

Notes

1. I rely in this text on the ideas of many friends of mine, especially Matthew Flisfeder and Todd MacGowan.
2. The film just extrapolates the tendency, which is already booming, of more and more perfect silicon dolls. See Bryan Appleyard, "Falling in Love with Sexbots," *The Sunday Times*, October 22 2017, pp. 24–25: "Sex robots may soon be here and up to 40% of men are interested in buying one. One-way love may be the only romance of the future." The reason for the power of this tendency is that it really brings nothing new: it merely actualizes the typical male procedure of reducing the real partner to a support of his fantasy.
3. I owe this point to Peter Strokin, Moscow.

4

Between the Capitalist and the Cop: The Path of Revolution in *Blade Runner 2049*

Todd McGowan

The Ideology of no Ideology

Most often, sequels are not necessary. Since Milton wrote *Paradise Regained* to capitalize on the unexpected success of *Paradise Lost*, writers, filmmakers, and studios have returned to profitable franchises to see if they can find some unexplored life in the corpse of the original. Since the composition of *The Odyssey*, the number of necessary sequels has exponentially dwarfed the number of contingent ones, sequels created for the sake of opportunistically exploiting a possibility rather than responding to new exigencies. But in 2017, a contestant for the exclusive category of the necessary sequel emerged, a possible newcomer to this pantheon that includes *2046* (Wong Kar Wai, 2005) and *Terminator 2* (James Cameron, 1991). This film takes the central concerns of the original and pushes them into the political situation that developed in the 35 years

T. McGowan (✉)
University of Vermont, Burlington, VT, USA
e-mail: Todd.McGowan@uvm.edu

C. Neill (ed.), *Lacanian Perspectives on Blade Runner 2049*,
The Palgrave Lacan Series,
https://doi.org/10.1007/978-3-030-56754-5_4

since the original's appearance. The driving need of this sequel derives from the dramatic transformation of the political situation. In this sense, the sequel is not opportunistic or even optional but necessary.[1]

Both the original *Blade Runner* (Ridley Scott, 1982) and the belated sequel *Blade Runner 2049* (Denis Villeneuve, 2017) reveal the trajectory from ideological submission to emancipation. The situation requires a sequel because ideology is no longer what it was when the original appeared. Ideological control has become much more effective, but at the same time, the opening for radical change has expanded. This is the paradoxical political situation that demands a new version of *Blade Runner*, one which Denis Villeneuve provides.

In the original, emancipation occurs just for a few individual replicants—including Rick Deckard (Harrison Ford) and Rachael (Sean Young). In the sequel, we again see the emancipation of an individual replicant, K (Ryan Gosling), but this film also shows K participating in a revolution of all replicants. Here, emancipation of one is a part of the emancipation of all. *Blade Runner 2049* envisions a possibility for revolutionary transformation that *Blade Runner* does not because it focuses on the contradiction between the forces of capitalist production and that of police power. The disjunction between these two axes of society opens up the space for the emergence of an alternative, represented in the film by the replicant revolution. Though *Blade Runner 2049* depicts a more effective form of ideological control than the original film does, it also recognizes the opportunity for a collective act once one breaks from this control, which the original cannot do. The original film presents a more vulnerable image of ideology, but it is only the sequel that can envision a collective challenge to social inequality.

The result of the change in the way that ideology functions between the two films is a fundamentally different conception of emancipation. Though the two films share a title and a mise-en-scène, they provide competing diagnoses of our subjectivity and of the subject's path to freedom. The earlier film depicts a typical understanding of ideology, one traceable to the thought of Louis Althusser. According to Althusser, ideology causes individuals to believe that they are their symbolic identity, that they are what the social order tells them that they are.[2] Here, ideology involves a fundamental duping. The individual sees itself as it isn't as a result of the operation of ideology—instantiated by the family, the educational system, the religious institution, the media apparatus, and so on. The objective correlative of ideological manipulation in Ridley Scott's *Blade Runner* is the implanted memories that the replicants have.

An implanted memory works akin to Althusser's ideological interpellation. It convinces the replicant that it has a symbolic identity that is not simply the result of the social structures that produce it. The implanted memory provides the patina of a free self for the replicant while actually serving as an index of the replicant's unfreedom. If the replicant has memories that it believes are its own, then it can believe in itself as a substantial being existing beyond social manipulation. It can believe itself to be a subject with mastery over its own existence. According to this conception of ideology, freedom depends paradoxically on recognizing one's unfreedom.

Though most of the replicants in the original film do not have implanted memories, they are all programmed to believe that they are nothing but their symbolic entity. In the case of those who have no implanted memories, this means that they believe that they actually are the role that they serve—laborer, pleasure provider, and so on. Though we never see the initial revolt of any of these replicants, the revolt depends on their ability to recognize that these identities are the result of manipulation and unfreedom.

In his analysis of the original *Blade Runner* in *Tarrying with the Negative*, Slavoj Žižek gives voice to this way of understanding both ideology and the film. He states, "it is only when, at the level of the enunciated content, I assume my replicant-status, that, at the level of enunciation, I become a truly human subject. 'I am a replicant' is the statement of the subject in its purest—the same as in Althusser's theory of ideology where the statement 'I am in ideology' is the only way for me to truly avoid the vicious circle of ideology" (Zizek, 1993; 41).[3] The discovery of one's status as a replicant marks the moment at which one becomes a free subject. At this moment, one ceases to confuse what Žižek calls the enunciated content of one's identity—one's personality, beliefs, memories, and attitudes—with one's subjectivity. Whereas identity results from one's historical and social situatedness, subjectivity emerges out of the ability to be indifferent to one's situation. In this sense, it is the site for the subject's freedom from all the factors that condition what the subject is. Recognizing that one's memories are an ideological deception frees the subject that makes this recognition.

Blade Runner depicts the emergence of freedom out of the ideological illusion of freedom. Only when Deckard and Rachael come to recognize that their memories are not their own do they gain the freedom that they previously believed that they had. In the conception of the original film, we attain freedom at the moment we discover our replicant status and take stock of the profundity of our immersion in the structures of ideology. The film also shows replicants without any false memories, like Roy Batty (Rutger Hauer) and Pris (Daryl Hannah), who are able to revolt because they have the form of subjectivity without any content that would tie them to their identity as slaves. They attain freedom when they confront the void of their identity.[4] Ideological control continues to function only as long as it works unawares. The moment that we know it is operative is the moment that we have the opportunity to break from its dominance.

This conception of freedom has its basis in a version of Althusser's theory of ideology. For Althusser, ideology depends on the existence of the unconscious, which is why it is not coincidental that at the moment he begins to introduce his definition of ideology he compares it with Freud's understanding of the unconscious. He writes that he is "proposing a theory of ideology *in general*, in the sense that Freud presented a theory of the unconscious *in general*" (Althusser, 1984; 35). Ideology works because the subject's self-division—the existence of the unconscious—makes it possible for the subject not to know what it is doing. If there were no unconscious, there would be no chance for the subject to be an unknowing dupe of ideological manipulation. The subject would have to submit consciously to the duping, which doesn't make any sense at all. This is why it took Freud—and Jacques Lacan—for Althusser's theory of ideology to become conceivable. Even though he used the term 'ideology,' because he wrote before Freud, Marx lacked the theoretical apparatus for making sense of his own concept.

Blade Runner 2049 paints a vastly different picture of ideology and freedom than the original, one in which the unconscious plays an even more prominent role. The sequel does not reject the emergence of freedom out of ideology that the original film depicts but shows how ideology has come to take this into account. In the universe of the sequel, newer replicants have ceased to revolt and become completely docile servants. The revolt that Roy Baty (Rutger Hauer) leads in the original has no correlative in the universe of the sequel as the film opens. Deckard's blade runner equivalent in *Blade Runner 2049*, K, is necessary to hunt down older replicants only. The newer models pose no revolutionary threat, no matter how much awareness they obtain.

If the newer film followed the model of the older one, we would imagine that the manufacturers of replicants perfected the art of duping them with their memories, that newer replicants don't revolt because they so thoroughly believe in their given symbolic identity that they never question it. But the film takes the opposite course. Rather than depicting replicants who believe completely in their memories (and thus in who they are), the film shows them not duped at all.

Even though replicants continue to have false memories and thus a sense of symbolic identity in *Blade Runner 2049*, they do not believe in these memories. The replicants experience the false memories as if they are real, but they know that they aren't. They seem to see perfectly through the ideological deception, in complete contrast to the replicants from the original film. This shift marks the later film's most important intervention into the functioning of ideology.

When K's commander Lieutenant Joshi (Robyn Wright) comes to K's apartment to order him to retire the replicant who has been born rather than created and thus represents a threat to the basic division between human and replicant, she asks him about his memories. He tells her, "I have memories, but they're not real. They're just implants." Despite this admission, Joshi wants to know the details. K responds, "I feel a little strange sharing a childhood story, considering I was never a child." K evinces complete awareness of the falsity of his memories. We see with this statement that ideology does not dupe him with the memories. This is a feature that K shares with all the new version of replicants—and none of them, unlike the older models, revolts. If ideology dupes K at all, it must do so with how he relates to the memories rather than with the memories themselves, unlike in the original *Blade Runner*. For this reason, this marks the crucial scene in the film for its conception of how ideology functions.

Earlier in the film, the first appearance of Joi also makes clear this conception of ideology. Joi (Anna da Armas), who is just a projection without any physical body, brings K his dinner after his day at work. Since Joi has no substantial existence, she cannot actually prepare K dinner. Instead, she brings a projection dinner of steak and French fries that she sets on the table in front of him. The film shows this dinner layered on top of K's real dinner, which is just a bowl of unappetizing gruel.[5] K has no doubt that Joi is not his real girlfriend or that his meal is not a bowl of gruel. Nonetheless, he acts as if they are real through his investment in the deception. If he weren't invested, he would simply eat the gruel without the projections of Joi and the tasty dinner.

Later in this same scene, K tells Joi that he has a present for her. She asks what the occasion is, and he answers, "Let's just say it's our anniversary." She wonders, "Is it?" He tells her that it isn't really but that they will just say that it is. Since Joi had no way of knowing whether it was their anniversary or not, K could have deceived her. But he opts to be straightforward—not to avoid the deception but to make the other aware of the deception. This is how ideology becomes even more effective. It works by laying its cards on the table.

Ideology necessarily deceives, but it works most effectively not when we believe in its deception but when we believe that we see through it. When we believe that we see through the ideological deception, the deception takes hold of us even more powerfully than when we remain completely duped by it. This is what prompts Jacques Lacan in his seminar entitled *Les non-dupes errent* to proclaim, "the non-dupes are two times dupes" (Lacan, unpublished). When one sees through the ideological deception, one eliminates the crucial moment for the emergence of subjectivity—the moment of doubt. The introduction of doubt coincides with the introduction of subjectivity in the modern epoch, from Shakespeare's decision to begin *Hamlet* with a question about the subject to Descartes' claim that the capacity for doubt provides the basis for the subject's freedom.[6]

One is not free when one gains certainty of the absence or the overcoming of ideological manipulation. One is free when one has the ability to doubt, which occurs in the moment that one questions one's ideological manipulation. At the point of doubt, the split between who the subject is and what the subject is becomes evident. The subject sees that it is not what it is. Only through the act of doubting can the subject emerge as the distance that doubt creates from the particular content of subjectivity. Doubt is alienation from identity, and identity, even if it is based on real memories rather than implants, is always ideological. By alienating the subject from what it thinks that it is, doubt locates subjectivity beyond the trap of identity. This is how it frees the subject. K gains this freedom in *Blade Runner 2049*, but he comes to it unknowingly.

K mistakenly associates freedom with being born rather than being created. As the film goes on, he attaches his belief in his own subjectivity to the possibility that he was the replicant born from a replicant mother rather than created in a laboratory. This becomes apparent in a conversation with Lieutenant Joshi after she orders him to track down and kill the replicant who was born. After K notes that he has never retired anything that had been born before, Joshi asks him why this matters. He responds, "To be born is to have a soul, I guess." This statement and K's subsequent investment in uncovering that he is the replicant who was born reveal K's theory of freedom: for K, freedom derives from birth because it indicates a starting point—a soul—that does not just result from manipulation. But the film undermines K's belief, even as it depicts him attaining freedom.

When K meets the revolutionary group of replicants, he learns that he is not the privileged replicant but one created just like all the others. Seeing his disappointment, the leader, Freysa (Hiam Abbass), asks him, "You imagined it was you?" She adds, "We all wish it was us. That's why we believe." When Freysa says this, the film cuts to K's memory of the toy horse, though the status of this memory undergoes an immediate shift. We now see it as an implanted memory, as an indication of K's ideological manipulation, rather than as an experience that he lived. Though this is an ideological fantasy, K's ability to doubt that it is just ideological is the key to his ability to join the revolution. The fact that K could imagine that his memory was real and doubt its inauthenticity provides the moment through which his freedom can emerge.

This sequence indicates that it is not the soul, as K initially believes, that is the source of freedom but subjectivity, which is the distance that one has from one's soul. If the subject were identical to the soul (which the film aligns with the fact of being born), what one was would be determinative for how one acts. The soul, in this sense, is just a natural determination that produces an animal being. This is why Aristotle considers the soul not what defines humanity but "in some sense the principle of animal life" (Aristotle, 1984; 641). This natural determinism would be indistinguishable from that of ideological manipulation functioning without a hitch. Nonetheless, as the revolutionary leader Freysa makes clear, subjects must be able to believe in the fantasy of a soul—they must doubt the efficaciousness of ideology—in order to discover their freedom. Those who are certain that they have been manipulated have been successfully manipulated. The elimination of doubt is the elimination of the margin for freedom.

The film illustrates the evacuation of the space for doubt through its visual and audible evacuation of the space for thinking. Even more than in the original film, characters in the world of *Blade Runner 2049* suffer from constant visual and audio bombardment. When K walks in the street, he encounters blaring advertisements, bright lights, precipitation, and whining machine noises. These define the mise-en-scène of the original *Blade Runner*, but they saturate *Blade Runner 2049* even more. At no point do we see K able to find any respite from the ceaseless noise and light. There are no visual or audible blank spaces in the world of the film. Subjects cannot doubt not only because they know that they have been manipulated but also because they have no opening in which to think.

The rewriting of ideological manipulation and the path to freedom that *Blade Runner 2049* enacts relative to *Blade Runner* marks a significant theoretical advance. But it also speaks to the contemporary age, an age when everyone recognizes the functioning of ideology and evinces awareness of its manipulations. As the film illustrates, ideology functions best when it operates openly. By doing so, ideology strips subjects of their capacity to doubt whether or not they have been manipulated. Once they know that they have been, they cede their capacity for freedom.

Despite the ramping up of the power of ideology that *Blade Runner 2049* depicts, it actually relates a more revolutionary situation that the earlier film. Ideology may have become more effective through the dissemination of knowledge about the manipulation that it performs, but at the same time a contradiction opens up between the force of capitalist production in the film and the symbolic authority. The primary opposition in the film is not between revolutionary subjects and the authorities or even between K and those opposed to him. It is rather the opposition between Niander Wallace (Jared Leto), the figure of untrammeled capitalism, and Lieutenant Joshi, the figure of the police. Freedom matters because it intervenes through the opening that this opposition creates. In this way, the film almost offers a practical guidebook for today's political situation.

What the Cop Wants

The revolutionary nature of capitalism stems not just from the incredible productivity that it unleashes on the world but from the disorientation that it introduces into the relationship between politics and economy. Traditional societies confine economy to the oikos or household, which is why Aristotle sees it as the domain fit for only slaves or women. With capitalism's emergence, economy, while never ceasing to remain a private arena, comes to dominate the political field. Economic interests begin to regulate political questions. Economy and politics lose their distinctiveness in the modern capitalist universe.

This loss of distinctiveness between economy and politics leads to the quagmires that beset leftist politics in modernity. Marx points out the contradictions of capitalist modernity in economic terms and sees politics as the superstructural response to these contradictions. When he envisions proletarian revolution as the crowning political act of modernity, he never theorizes exactly how this comes about or what type of political act is necessary. For Marx, politics follows from economics, which is why he leaves politics so little theorized. Economic contradictions create the opening for political acts that take advantage of these contradictions, but politics alone cannot change the economic terrain. If the time is not ripe for revolution, no political act could possibly be successful, although the political act must always anticipate the proper time rather than waiting for it. The revolt of Spartacus could not lead to a communist society because the means of production had not yet paved the way for it.

Theorists in the twentieth century spend much time freeing themselves from Marx's tendency to give politics a secondary place relative to economics. From Hannah Arendt to Frantz Fanon to Alain Badiou, the important theorists of the twentieth century stress that the political act has the ability to change the symbolic terrain and thereby alter the economic situation. Politics, in their view, must at least be capable of transcending economics.[7] Arendt, for her part, sees Marx and capitalism as peas in a pod for their shared devaluation of the political. Alain Badiou doesn't go this far, but he privileges politics to such an extent that he refuses to allow for an economic event in his theory of the event. As Badiou sees it, politics is a truth procedure and thus can produce revolutionary events, but economics cannot possibly be. There is no economic revolution; there is instead a revolution that frees us from the dominance of the economic and its insistence on number.[8]

A leftist opposition develops between Marxist challenges to capitalism on the one hand and political challenges to the ruling symbolic structure on the other. The leftist doxa that stems from Marx himself interprets these two forces—economics and politics—as manifestations of the same logic. To fight against capitalism is to fight against symbolic authority, and to fight against symbolic authority is to fight against capitalism. *Blade Runner 2049* calls this transitive property of struggle into question in the most thoroughgoing way possible.

The struggle between Wallace, the representative of unbridled capitalist productivity, and Joshi, the figure of the police, drives the action of the film and opens up the revolutionary possibility that the replicant revolution seizes. As in the original, the police that we see in the sequel have as their task guarding the barrier between human and replicant. This barrier exists in order to foment capitalist productivity. By ensuring that replicants remain a distinct order of beings without the freedom that humans have, the police maintain them as a readily available source of labor for the capitalists. In this sense, capitalism requires the police to fulfill a function that sustains capitalist productivity.

Without the barrier between human and replicant, all would view themselves as equal. Replicants would refuse to continue as exploited labor. This leads Police Lieutenant Joshi to fear the discovery of a replicant who was born and to send K on a mission to eliminate all trace of this replicant's existence. After K discovers the existence of this replicant, she tells him, "The world is built on a wall that separates kind. Tell either side that there's no wall, you bought a war—or a slaughter." Joshi senses the danger that a replicant who was born would pose to the societal order. Her task, as she sees it, is to defend this order from any possible disruption.

As Joshi articulates her understanding of the role of the police—"we keep order," she tells K—the film takes a less sanguine view of that order than Joshi herself. Her conversation with K begins with an external shot through the window of her office. This shot highlights the steady rain that defines the world of the film (until K later goes to the apocalyptic wasteland of Las Vegas). By initially showing Joshi's panegyric to order through a shot depicting the dismal external world, what she calls order appears as disorder. Then, as K leaves the police station, we see an enormous advertisement that proclaims, "Enjoy Coca-Cola" out in the rain. Joshi defends the supposed order of a society going out of control in the thrall of a rapacious capitalism that transgresses every limit.

Joshi sends K in search of the replicant who was born in order to eliminate this threat to the ontological division that sustain the prevailing social structure. She understands that mass awareness of the birth of a replicant would have the effect of reducing the ontological difference between replicant and human to the class divide. This is the great fear of all conservatives. If we recognize that ethnic difference exists to make possible class exploitation by obscuring class division, class society becomes impossible. The inequalities of class society are tolerable to us because we attribute them to ontological difference—difference in race, in sex, in genetic predisposition, and so on. The moment that we cease investing ourselves in some form of ontological difference, the justification for class inequalities disappears. This is what Joshi understands and why she believes in the necessity of wiping out the idea of a replicant birth.

Joshi ultimately fails in her attempt to hide the birth. But her failure is not the result of the group of revolutionary replicants led by Freysa that we see gathered near the end of the film. Instead, it is the capitalist society that she aims to safeguard that both thwarts her plan and kills her. Wallace's replicant assistant Luv (Sylvia Hoeks) sneaks into the police station to steal Rachael's remains (which indicate the birth of a child) and kills Joshi's assistant Coco (David Dastmalchian). Later, Luv returns to the police station and stabs Joshi in her office. She does so at the behest of Wallace, who orders her to find the replicant who was born. These two murders demonstrate clearly that the police and the capitalist are not on the same side, despite the fact that the police does the work that the functioning of capitalism requires. The capitalist needs the police, but the police acts as a barrier to capitalist productivity that the capitalist would like to remove. The capitalist both depends on the barrier that the police creates and finds this barrier an intolerable obstacle that it must extirpate. This is one of the basic contradictions that besets capitalist society.

This situation depicted in the film sheds light on the relationship between the capitalist and undocumented immigrant labor in our contemporary universe. While on the one hand the capitalist needs the division between citizens (or documented immigrants) and undocumented immigrants to obscure the class divide, on the other hand this division acts as a barrier to capitalist production. In the capitalist's dream world, an influx of undocumented laborers would make it possible to expand dramatically the exploitation of labor and thus profitability. Documented laborers cost more than their undocumented counterparts, which is why every good capitalist would prefer to hire the undocumented. In this sense, the capitalist, like Wallace in the film, is the enemy of the border patrol, represented by Joshi with her desire to sustain a wall between kinds. Even though the police help to sustain the division that the capitalist requires, this division costs the capitalist in profit, which opens up a conflict between the capitalist and the police.

What the Capitalist Wants

It is clear from early on in *Blade Runner 2049* that the figure of capitalism in the film sees his enemy not so much in potential revolutionaries but in the police. When Wallace hears about the birth of a replicant, he begins to talk in terms of Christ's birth. He says to his assistant Luv, "Can you at least pronounce 'a child is born'?" But it is not initially clear when Wallace says this that he is talking about Rachael's child or about the new replicant model just produced. He says to Luv, "A new model—well, let us see her then." The film cuts to a newly created replicant dropping naked from the ceiling. When Wallace goes to examine the new model, the ambiguity of what he is talking about remains, though it soon becomes evident that he knows about the childbirth and that he intends to make use of it for unlimited labor power.

As Wallace examines the newly minted replicant, he describes his plan. He outlines a project of massive colonization of space, proclaiming grandiloquently, "We should own the stars." He sees in replicant childbirth the path to this possibility. He explains, "Every leap of civilization was built on the back of a disposable work force. But we lost our stomach for slaves, unless engineered." With the development of replicant childbirth, Wallace could expand the number of replicants exponentially—much faster than if he just manufactured them—and thus create the "disposable work force" that his dream requires. He believes that with this labor power, "We could storm Eden and retake her." This dream of capturing paradise lost is the dream that animates all capitalist accumulation, though none has probably ever articulated it quite so overtly as Wallace does in this scene.

While Wallace announces his plan, he sits in a chair a few steps in front of the new replicant. Because he is blind, he surveys her through multiple small drones that circle her body under his mental control. The film highlights Wallace's unusual form of inspection in order to indicate his removal from labor. Wallace designs replicants but subtracts himself from all labor processes. His dream of the conquest of the stars is a dream that must have nothing to do with any laboring on his part because labor provides a constant reminder that one remains outside of paradise and continues to lack.

The energy for capitalist accumulation comes directly from the promise of a future abundance that would bring with it complete satisfaction. This is what Wallace alludes to when he mentions Eden during the explanation of his plan for the conquest of space. He believes that a total conquest would bring total satisfaction. There would be, at this point, nothing left to accumulate, which is the dream that drives capitalism.

But the vastness of space gives the lie to this dream: even a work force of breeding replicants would never be able to colonize the entire universe. No matter how much Wallace conquered, there would always be more to go. The dream that energizes capitalist productivity does so only insofar as it remains constitutively unfulfilled. The capitalist such as Wallace doesn't want to realize this dream but instead finds enjoyment elsewhere.[9]

The film makes this evident in the radical divide between the visual and the audio track during this scene. As he initially surveys the replicant, Wallace evinces apparent sympathy for her. But his actions belie any suggestion of sympathy. After proclaiming his dream for intergalactic expansion relying on replicant labor, he laments, "But I can only make so many." Just before he says this, the film cuts to a close-up of Wallace's hands fondling a knife that he will use to kill the newly minted replicant standing naked in front of him. He talks about creation while enacting destruction. After slicing her abdomen with the knife, Wallace states, "We need more replicants than can ever be assembled—millions, so we can be trillions more." At the precise moment that he dreams of infinite creation, Wallace destroys the replicant standing in front of him. Though he notices the replicant's absolute fear, he not only fails to evince compassion for her but wantonly destroys her. The film includes this gratuitous killing on his part to highlight the underside of his dream. While it seems as if the dream of recapturing Eden motivates Wallace, his enjoyment resides in the unnecessary destruction that pursuing this dream entails. This is the enjoyment that feeds the capitalist subject and gives the lie to capitalism's vision of the future.

But the expansiveness of Wallace's dream is also the key to its ability to unleash unwittingly the forces of revolution. He orders Luv to assist K in tracking down the missing replicant child. Her help proves instrumental in enabling K to survive when the police or other forces try to stop him.

In the end, K is able to join the forces of the replicant revolution due in large part to this assistance. Wallace's dream places him at odds with the police and thereby opens up the space through which a revolutionary change to the society can be introduced through those excluded both politically and economically.

The forces of capital (like Wallace) constantly drive to expand themselves in a way that puts them in conflict with the social structure that initially nourishes these forces. In this way, capitalists strive to undo the barrier that is necessary for the capitalist system. It survives because their efforts cannot succeed. This is why the forces of capital and the police are not simply different expressions of the same logic. Instead, they only align incidentally and are for the most part at odds with each other. Emancipatory transformation occurs not through combatting one or the other or both at the same time but through taking advantage of their conflict. This is the path toward emancipation that *Blade Runner 2049* highlights.

From Noir to Anti-noir

Emancipation in *Blade Runner 2049* goes through K, who goes from being a docile replicant wiping out other replicants to being a self-sacrificing participant in revolutionary struggle. The transformation of K mirrors the transformation that Deckard undergoes at the very end of *Blade Runner* when he decides to run away with Rachael. But Deckard does not undergo the radical awakening that occurs with K. The replicant Roy Baty (Rutger Hauer) displays more awareness of the situation of the replicant than Deckard in their final fight. Whatever awareness Deckard has, he has it only begrudgingly. This distinguishes him from K, though both reveal how freedom can emerge out of an apparently invincible ideological control, which is the problem that defines film noir.[10]

The original *Blade Runner* film is epochal not least for its introduction the style and structure of neo-noir into science fiction. Ridley Scott creates a film in which the dreary mise-en-scène of a future Los Angeles matches that of John Huston's *The Maltese Falcon* (1941) or Billy

Wilder's *Double Indemnity* (1944).[11] The steady rain that falls conveys the noir atmosphere, but even more, the narrative trajectory signals the concerns of classic film noir. Though the bounty hunter Rick Deckard is nominally pursuing renegade replicants, he is more importantly, albeit unbeknownst to himself, investigating his own identity. The trajectory of film noir—and this the primary feature that separates it from stories of classical detection—is always one that folds back on itself. As the hero's investigation progresses, it always ends up involving the hero and the hero's own status as a subject.

The locus classicus for this procedure is John Farrow's *The Big Clock* (1949). In this film, the noir hero George Stroud (Ray Milland) conducts an investigation on behalf of his boss, Earl Janoth (Charles Laughton), whose real target is Stroud himself. As a result, Stroud must delay and obstruct the investigation without arousing his boss's suspicion that he is the one Janoth is trying to find. If Stroud does not successfully impede the search, he will find himself implicated in the murder of Pauline York (Rita Johnson), a woman he spent time with during an evening but whom Janoth later killed. Janoth wants Stroud to help him pin the murder on Stroud, though he doesn't know that Stroud is the mystery man who was with York. The closer that Stroud comes to the solution, the closer he comes to his own undoing.[12]

A similar dynamic unfolds in *Blade Runner*, though Deckard, unlike Stroud, remains unaware of his own status throughout the film. When he pursues and kills the replicants, Deckard unknowingly pursues and eliminates his own comrades. He never fully catches on to his connection with the replicants despite the numerous clues that we see placed for him, including Rachael's question about whether he ever underwent the replicant test himself and the origami unicorn left on the floor of his apartment at the end of the film. Even though Deckard's love for Rachael evinces his awareness of a connection with the replicants, he does not display any class solidarity.

Blade Runner 2049 sets up the spectator for the noir narrative structure of the original film. When K hears of the miraculous birth from a replicant mother, clues quickly begin to mount that suggest K himself is the replicant who was born rather than created. These clues begin when K first discovers Rachael's grave at the farm of Sapper Morton (Dave

Bautista). K's fantasy object, the black horse, has the date inscribed on it that he discovers carved into the tree just above the grave: 6-10-21. He begins to believe that this date of the first replicant birth is actually his own birthday and that he is this replicant.

If K were the one replicant that was born—if Deckard were K's father—the film would be a film noir in the tradition of the original. In the noir universe, the narrative loop always closes back on itself. The subject discovers what it always already was.[13] The trajectory in *Blade Runner 2049* veers off before reaching this point. In this film, K discovers what he isn't. Instead of recognizing what he didn't know about himself, K finds out that he isn't what he thought he was.

This separates K from his namesake in Franz Kafka's *The Trial*. The obvious allusion to Franz Kafka's *The Trial*—made especially evident when Joi nicknames K "Joe" and makes unmistakable the link to Joseph K—has the effect not of aligning the two characters but of emphasizing the position that Villeneuve's K abandons. Initially, he is like Kafka's Joesph K—completely certain of himself and what he is. As the film goes on, however, Villeneuve's K acquires the ability to doubt in a way that Kafka's K does not. Kafka's K goes astray because he never doubts his own identity as an individual despite the assault by figures of the law, whereas Villeneuve's K does begin to doubt. In this way, Villeneuve's K attains a freedom that Kafka's K does not.

Perhaps the most conspicuous difference between K and his bounty hunter forerunner is his fantasy object. For Deckard, as we see clearly only in the director's versions of the film, his fantasy object is a unicorn. As he solemnly presses the keys of the piano alone in his apartment and looks at old photographs, the film cuts between a close-ups of Deckard's face looking absently and two shots of a unicorn running through a woods. This sequence establishes the unicorn as Deckard's fantasy object, a status confirmed later when his fellow police officer Gaff (Edward James Olmos) leaves a origami unicorn on the floor of his apartment. This origami unicorn suggests Gaff knows Deckard's fantasy object, which would only be possible if this object, like the memories of the replicants in the film, was implanted. Gaff's clue to Deckard is also the most telling clue for the spectator that Deckard himself is not simply pursuing replicants but actually is one himself.

In *Blade Runner 2049*, K has a fantasy object that plays an even more prominent role than Deckard's unicorn. It also provides the key to understanding K's existential status, although K, in contrast to Deckard, never has any doubt that he is a replicant. Whereas Deckard's fantasy object is an imaginary animal, K's is a real one. K has a horse that corresponds to Deckard's unicorn. What's more, K actually finds the toy horse that corresponds to his memory, whereas Deckard obviously never encounters an actual unicorn. The reality of the object leads K to believe it represents a real memory rather than an implanted one.

After he finds the toy horse hidden at an orphanage that corresponds to his memory, K goes to see the foremost creator of memories—Dr. Ana Stelline (Carla Juri). She examines the memory and relays to K, ambiguously, that someone actually lived this memory. What she doesn't tell him (or the spectator) is that she was this someone, not K. The memory was one of her own real memories acquired when she was at the orphanage as a young girl. When K finally realizes that he is not the privileged replicant, he comes to understand that the reality of the memory and the object can be the source of the greatest deception. Reality hides the truth of subjectivity even better than fantasy.

In the original film, the fantasy object reveals the truth about Deckard's subjectivity. In the sequel, the fantasy object becomes a real object but nonetheless deceives K about the nature of his subjectivity. The reality of the object that seems to verify the truth of the fantasy actually ensures the deception about K's subjectivity. K discovers who he is only when he recognizes the false status of the real object, while Deckard discovers who he is when he recognizes the true status of the fantasized object. Both subjects find the truth of their subjectivity, but they do so in opposite ways because they confront different ideological structures obscuring their subjectivity in opposite ways.

Pale Fire

The shift in the functioning of ideology from the original film to *Blade Runner 2049* has its objective correlative in the test that we see replicants taking in each film. In the original, blade runners administer the

Voigt-Kampff Test to would-be replicants in order to detect the lack of an emotional response that would indicate replicant status. In the sequel, the police give blade runners a baseline test to ensure that they are not developing emotional attachments. Though the tests play opposite roles, their ultimate function is the same in both films—to sustain the clear dividing line between human and replicant, between those who belong to the society and those who provide labor for it.

After returning from his mission to kill Sapper Morton, K submits to his baseline test. He takes the test in an all-white room where he faces a screen that displays the words that K must repeat after a computer voice speaks them to him. The test occurs in a room in the police station, but we don't see a connection between this room and the rest of the station. Villeneuve overlaps the audio of the scene beginning with the visual of K flying toward the police station in his ship before cutting to K in the room. When the test ends and the voice administering the test tells K to pick up his bonus, the film cuts directly to K walking through the street outside. There are no other internal shots of the police station that would link the testing room to the rest of the police station. The depiction of the test reveals K as a completely isolated subject that nonetheless has no opening for his singularity. This isolation mirrors K's own isolation when he arrives home to jeers from those who live in his apartment building and to the words "fuck off skinner" written on the door to his apartment.

The significance of the baseline test that K undergoes in the scene derives not just from how Villeneuve constructs the scene but even more from what K actually says. Before K appears seated in the room, we hear the male voice of authority tell him, "Officer KD3-6.7 … recite your baseline." K responds by reciting lines from Vladimir Nabokov's novel *Pale Fire*.[14] These lines articulate K's ideological situation, a situation that follows from the visual structure of this scene. Quoting directly from the novel, K states, "And blood-black nothingness began to spin / A system of cells interlinked within / Cells interlinked within cells interlinked / Within one stem. And dreadfully distinct / Against the dark, a tall white fountain played" (Nabokov, 1962; 59). K views himself as an individual cell isolated from all others, but at the same time, he finds himself interlinked within his isolation. He is distinct and yet has no distinctiveness because he remains trapped within the interlinked identity given to him by the social authorities.

When K finishes this quotation from Nabokov, the voice of authority asks K a series of questions. After each question that the voice asks, it states, "Interlinked." Rather than answer any of the questions, K repeats this word, which suggests that the ideological mechanism does not treat him as a subject but as a machine. He responds by affirming his connection to the whole rather than by expressing his desire through an answer. This is crucial for how ideological interpellation works since it must assure the subject that it is its identity, a process that the assertion of subjectivity undermines.

The wall in front of him initially appears from behind his head as a white blur, but as the camera tracks forward closer to his head, the black camera surveying him on the front wall of the room comes into focus. The camera finally comes into complete focus when the voice asks, "Do you feel that there's a part of you that's missing?" This question stands out from the others because it provides the mechanism for ideological control. Ideological fantasy promises an object that would be the part of the subject that is missing. It holds out this promise as a way of attracting the subject's libidinal investment.

The test ends when the voice asks him to repeat three times, "Within cells interlinked," a phrase that emphasizes K's ideologically determined isolation and determined connection to the structure within that isolation.[15] While the citation provides a succinct description of K's situation, its role in Nabokov's novel provides the key to the ideological power of the baseline test. The novel consists of a foreword and commentary surrounding a 999-line poem. The fictional author of the foreword and commentary is Charles Kinbote, while the author of the poem is a character named John Shade. The action of the novel occurs primarily in the commentary, which recounts Kinbote's escape from the country of Zembla where he was a king deposed by a revolution. The central mystery of the novel concerns the death of John Shade. We learn about his death in the foreword but don't uncover the circumstances until the end of the commentary. And even then, there are several clues that we shouldn't trust Kinbote as a narrator—because he is likely insane—which casts doubt on his contention that an assassin trying to murder him mistakenly killed Shade.

Just as the circumstances of Shade's death is the central mystery in the novel, the key enigma in the poem itself also involves death. Shade has a near-death experience when he suffers a heart attack during the question and answer period of a talk on poetry he is giving. As he collapses, he has a vision in which he articulates the lines that K quotes during his test. Prior to this experience, the problem of what happens to us after death preoccupies Shade. He devotes himself to "the truth / About survival after death," which he assumes that everyone knows but him (Nabokov, 1962; 39). The moment of his collapse provides him insight not only into the ultimate question of human existence but also into this question that most vexes him personally.

The vision of the white foundation that comes to Shade provides him a definite answer. After this vision, he comports himself differently. He states, "There in the background of my soul it stood, / Old Faithful! And it's presence always would / Console me wonderfully" (Nabokov, 1962; 60). Shade finds consolation in the certainty of the white fountain, which has the certainty of Old Faithful. Shade finds further confirmation of the objective truth of his vision when he reads about a woman who suffered a heart attack and also had a vision of a white fountain. When he learns about this woman, Shade feels confident that the fountain was not just his own fantasy but was a real object of the next world.

But from the journalist who wrote the article about the woman's experience Shade learns that the magazine included a misprint. The woman reported seeing a white *mountain* rather than a white *fountain*. Shade is initially disappointed and writes, "Life Everlasting—based on a misprint!" (Nabokov, 1962; 62). This disappointment occasions a revolution in Shade's understanding of his vision. Instead of believing that he found the only possible answer, he comes to see that the solution lies solely in the groping toward a solution, not in a specific answer. The confrontation with the misprint breaks up Shade's certainty and reintroduces doubt into his existence.

It is not a coincidence that when K returns from a mission he must recite the lines from John Shade's vision in *Pale Fire*. Just as these lines indicate Shade's discovery of absolute certainty, K's baseline test attempts to verify that he has not begun to doubt—that is, that he has not yet begun to be a subject. The lines from *Pale Fire* suggest a clear answer to the ultimate existential question—knowledge about what happens to the individual after death. Until his attempt at verification goes awry, these lines mark a moment of certainty for Shade. If K can repeat them without emotion, this tells the authorities that he accepts the certainty of what he is and his place within the society.

The test scene combines the mise-en-scène of different films—the overwhelming white color of *THX 1138* (George Lucas, 1971) and the setting of *A Clockwork Orange* (Stanley Kubrick, 1971) and *The Parallax View* (Alan Pakula, 1974). These latter two films show an individual being programmed by a series of images projected in front of him. While *A Clockwork Orange* and *The Parallax View* depict the individual as a passive recipient of the images, *Blade Runner 2049* shows K repeating the words spoken to him. Instead of manipulating him, the words attempt to guarantee that K has not escaped from the psychical control that governs replicant existence. The attempt at manipulation of the earlier two films becomes securing the fact of manipulation in the later one.

Unlike *A Clockwork Orange* and *The Parallax View*, the test in *Blade Runner 2049* does not, as we might expect, bombard K with any images. The role of the test is different in the latter than in these earlier films. Rather than trying to program K, the police are trying to maintain his ideological programming. The absence of images from the authority confronts K with the pure voice.

Authority typically operates through a combined visual and audio approach to the subject. Authority gives the subject images to situate itself relative to and a voice that urges the subject to act. Typically, the visual field provides the ego ideal for the subject while the audio field gives the subject the model for its superego. The ego ideal is the image for whose look the subject tries to gain recognition. The superego is a voice that constantly impels the subject forward. Because his ideological interpellation is so successful, K requires no ego ideal. His inability to doubt keeps him confined within the prescripts of ideology. All he needs is the superego to drive him to act. This is why there are, somewhat surprisingly, no images in this test.

This scene of the baseline test that occurs approximately ten minutes into the film acts as a synecdoche for the ideological control that the film displays. K affirms his immersion within ideology by articulating his complete certainty through the passage from *Pale Fire*. K's recitation of this passage reflects the effectiveness of ideological control for contemporary subjects. Rather than controlling K through limiting his knowledge, ideology here works by giving K a sense of knowing the ultimate mysteries of existence. Believing that one knows all there is to know leaves one unable to engage in the fundamental act of subjectivity—doubting.

The Less You Are, the More You Are

In *The Human Condition*, Hannah Arendt worries what would happen to freedom if human beings left Earth and migrated to another planet. If this occurred, Arendt argues, we would lose the opening for breaking from our ideological conditions. We would become our own product and thus be incapable of genuinely acting to create novelty.[16] Because Arendt doesn't think of replicants as subjects, she never entertains this question in terms of them. But the situation for replicants and other technologically produced entities is even worse than for humans who have left Earth. They seem to be utterly lacking the opening through which the subject can assert its freedom because they have not been born (which is perhaps why Arendt never broaches the question of the replicant's freedom).

When K tells Joshi that the birth of a replicant changes the situation because such a being would have a soul, he expresses his devotion to Arendt's identification of freedom with the moment of birth. For Arendt, the fact of birth is integral to the possibility of freedom. A being that was not born cannot but follow from the dictates of its construction. But neither K nor Arendt have the last word on freedom in *Blade Runner 2049*. The film shows that it is not birth that creates the singularity of the subject in the film but the relationship to death. This occurs in the case of both K and Joi, his projection girlfriend.

The subject asserts its freedom through how it chooses to destroy itself. Both K and Joi assert their freedom in the film. They do so through a suicidal gesture that lifts them out of their determinations. In this way, the film takes Hegel's path instead of Arendt's. For Hegel, it is the capacity for freely limiting itself—for freely taking up its own form of self-destruction—that distinguishes the subject from the animal. In the *Philosophy of Right*, Hegel states, "The human being alone is able to abandon all things, even his own life: he can commit suicide. The animal cannot do this; it always remains only negative, in a determination which is alien to it and to which it merely grows accustomed" (Hegel, 1991; 38). One breaks out of a negative external determination only through the act of determining oneself, which is the act of limiting oneself. K does this when he decides to sacrifice himself for the revolutionary cause, and Joi does it when she chooses to leave the safety of K's apartment and risk her entire existence.

Blade Runner 2049 insists again and again that Joi is nothing but a fantasy object. When she first appears outside K's apartment on the roof of the building, she is shrouded in a light green glow just like at the moment in *Vertigo* (Alfred Hitchcock, 1958) when Judy (Kim Novak) becomes Madeline, the fantasy object of Scottie (James Stewart).[17] As Judy emerges from the bathroom after having made herself up to exactly resemble Madeline, Hitchcock bathes her in a green glow in order to make clear her fantasmatic status. The key to this fantasy is that she imitates someone who never existed: Judy acted the part of Madeline in order to seduce Scottie, so that there is no real Madeline that Scottie initially lost. At this moment, Judy is the paradigmatic fantasy object because she represents the return of an object that never existed in the first place.

This is why the green shading that accompanies Joi's appearance outside the apartment in *Blade Runner 2049* is significant. Like Judy, Joi replaces an object for K that never existed. He proclaims his love for her, and yet she is a commodity produced for his satisfaction, which is why we see a massive advertisement for her outside just after K leaves her. Though Joi is a pure product, the result of a manufacturing process, she nonetheless emerges as a subject in the film. Like K, she has no identity that doesn't derive from her programming. In fact, she is even more unsubstantial than K because she lacks any physical presence. Her subjectivity is purely fantasmatic. But the film shows that even this fantasmatic entity with no physical body can be a subject because it has the ability to sacrifice itself. When K takes her along with him as he flees from the police, he must download her entire program into the portable emanator. This puts her at extreme risk because the destruction of the emanator would entail her total destruction. Nonetheless, she agrees to go along, knowing that it is likely that total annihilation will ensue for her. Though Luv later destroys the emanator and Joi, Joi nonetheless asserts her subjectivity in the act of accepting this destruction.

Joi's act parallels K's, as he decides to sacrifice his life for the replicant revolution. Though he feels disappointment when he learns that he has not been born and is thus just an ordinary replicant, he finds recompense for this disappointment through his political act. As he lies dying at the end of the film after having saved Deckard's life and having brought Deckard to see his daughter, K looks down on the snow landing on his hand. At this moment, he exists as a free subject. This existence points to the possibility of freedom in the face of the most recalcitrant ideological structures.

Blade Runner 2049 depicts a world where ideological control has become more certain than ever. Rather than simply controlling subjects with implanted memories, the society lets subjects know about the falseness of everything they remember so that they have no capacity to doubt that they have been manipulated. The elimination of doubt does not take place without a hitch, however. It is this hitch, which occurs through the manufacture of replicants capable of reproduction, that marks the possibility of freeing subjectivity from its ideological constraints. At the moment ideology functions most effectively, the conflict between capital and the police creates a moment of emancipatory possibility. If *Blade Runner 2049* goes further than the original film in exploring the basis of this emancipatory possibility, it does so because ideology has become so much more effective in the intervening years. Freedom is only available when we recognize that our knowledge of how we've been manipulated is central to the fact of that manipulation.

Freedom does arrive when we completely overcome our ideological manipulation. There is no absolute escape from ideology. Even if there were, that would not be the path to freedom. Freedom depends on our ability to entertain the possibility that our ideological manipulation is not just ideological manipulation but actually true. Without the possibility that ideology is true, we lose our ability to act on its falsity because we lose the capacity to doubt. Who the subject is as a free subject emerges when the subject doubts what it is. This doubt only occurs when we cease relating to ideology with complete cynical distance, as K initially does in the film. *Blade Runner 2049* responds to our contemporary ideological situation by laying out a path to freedom radically different from the original film. At the moment of ideology's apparently total victory, its vulnerability consists in our renewed ability to believe in it.

Notes

1. The category of the necessary sequel begins with *The Odyssey*, though it isn't clear that it was composed after *The Iliad*. This is why works such as *2046* and *Terminator 2* fit this category even better than its inaugural work. What characterizes a contingent sequel is not only its failure to break new ground—*Return of the Jedi* (Richard Marquand, 1983) rehearses the same struggle as *Star Wars* (George Lucas, 1977) and even concludes with the destruction of the same weapon—but a political repetition as well. The populist rebellion against the evil empire functions in precisely the same way in both films, just as the infinite sequels in horror franchises replay the same political dynamic again and again. While one can easily name ten film sequels that surpass the originals and thus might qualify as necessary, other than *The Odyssey* it is difficult to think of many literary examples. Works like *Huck Finn* and *Ulysses*, while they seem to qualify because they employ the same characters as *Tom Sawyer* and *Portrait of the Artist as a Young Man*, are actually hard to characterize as sequels because the only thing that they have in common with the earlier novels is the characters. (I am indebted to many members of the English Department at the University of Vermont—including Elizabeth Fenton, Daniel Fogel, Huck Gutman, Major Jackson, David Jenemann, Mary Lou Kete, Lokangaka Losambe, Deborah Noel, Holly Painter, Tom Simone, and Sean Witters—for their suggestions about the status of various sequels).

2. According to Althusser, "Ideology represents the imaginary relationship of individuals to their real conditions of existence" (Althusser, 36).
3. Despite here referring approvingly to Althusser's theory of ideology, Žižek is actually one of the foremost opponents of this theory. For Žižek, subjectivity is not the result of ideological interpellation but the basis for rejecting the interpellation.
4. This becomes clear when Pris tells J. F. Sebastian (William Sanderson), "I think, Sebastian, therefore I am." Pris's citation of Descartes locates even the replicants that don't have any implanted memories in the Cartesian project of radical doubt that marks the subject's freedom.
5. This is a direct quotation of a similar scene in Terry Gilliam's *Brazil* (1985), which uses the same opposition between the appealing image and the unappealing actual meal.
6. The great philosophical breakthrough that occurs with Descartes concerns his move to base any certainty that he can discover on a prior moment of doubt. He shows that doubt is the fundamental gesture of subjectivity. When he constructs a system of certainty through the existence of God, he turns away from subjectivity itself and abandons the radicality of the initial insight. But we should not let this turn away color our appreciation for the immensity of Descartes' discovery.
7. Neither Arendt nor Fanon nor Badiou would appreciate being grouped with the others, but they do share a combined emphasis on the ability of politics to transcend economics that separates them from Marx, even at the points where they associate themselves with him.
8. In *Number and Numbers*, Badiou links the dominance of number to capitalism. He writes, "In our situation, that of Capital, the reign of number is ... the reign of the unthought slavery of numericality itself. Number, which, so it is claimed, underlies everything of value, is in actual fact a proscription against any thinking of number itself. Number operates at that obscure point where the situation constitutes its law; obscure through its being at once sovereign and subtracted from all thought, and even from every investigation that orients itself towards some truth" (Badiou, 2008; 213) At the same time that economics reduces all values to number, it creates an impossibility for subjects to theorize number, so that it can dominate without being questioned.
9. For a more detailed explanation of the capitalist enjoyment, see (McGowan, 2016).
10. This is what Robert Pippin is getting at when he notes, "film noir directors ... show us what it literally *looks like*, what it *feels like*, to live in a world

where the experience of our own agency has begun to shift." According to Pippin, film noir is an exploration of our potential for agency in a world that determines us completely.

11. Joan Copjec points out that the unending rain that often characterizes noir is the result of the insistence on one's private enjoyment without any regard for the public realm. She writes, "from the moment the choice of private enjoyment *over* community is made, one's privacy ceases to be something one supposes as veiled from prying eyes (so that, as in the case with Keyes, no one can be sure that one even has a private life) and becomes instead something one visibly endures—like an unending, discomfiting rain. In film noir privacy establishes itself as the rule, not as a clandestine exception. This changes the very character of privacy and, indeed, of 'society' in general—which begins with the introduction of this new mode of being to shatter into incommensurable fragments" (Copjec, 1994; 183). Copjec's description of the noir universe applies equally to that of *Blade Runner*.
12. This logic becomes even more explicit in the neo-noir remake of *The Big Clock*, *No Way Out* (Roger Donaldson, 1987), which stars Sean Young from the original *Blade Runner*.
13. Film noir develops the logic of Freud's *Nachträglichkeit* or belatedness to its endpoint. In the noir narrative, the hero's later actions have the effect of retroactively revealing what the hero was at the beginning. This is why film noir so often has recourse to flashback. The use of flashback suggests the temporality of belatedness, in which the future has the ability to reshape the past.
14. The novel first appears in the film when Joi picks it up in K's apartment and asks him to read to her. He responds, "You hate that book." She immediately tosses it aside and says, "I don't want to read either." We might surmise that Joi's hatred of Nabokov's novel derives from the fact that the central events of the novel end up being just the narrator's delusion. In this sense, Joi sees too much of K in the narrator Charles Kinbote. But this also explains why K would be interested in the novel and have it in his apartment.
15. The phrase "within cells interlinked" provides a near-perfect three word account of the contemporary situation for the subject. Subjects exist in isolation from others and at the same time find themselves interlinked through various social structures, from Facebook to Twitter to Tinder. Rather than eliminating the isolation, these interlinks have the effect of reinforcing it.

16. Arendt states, "The most radical change in the human condition we can imagine would be an emigration of men from the earth to some other planet" (Arendt, 1998; 10). Arendt links our capacity for novelty to the role that our interaction with Earth plays in our subjectivity. Without this contingent element, we become wholly determined.
17. When K arrives at his apartment building prior to this scene with Joi, the film depicts the staircase that he must climb to his apartment in a way that echoes the famous staircase shot from *Vertigo*. It is as if Villeneuve includes this shot of the staircase to prepare the spectator for the more significant allusion to *Vertigo* that would follow.

References

Althusser, L. (1984) Ideology and Ideological State Apparatuses (Notes Toward an Investigation), trans. Ben Brewster, in Essays on Ideology. London. Verso.

Arendt, H. (1998) The Human Condition, 2nd ed. Chicago. University of Chicago Press.

Aristotle (1984) On the Soul, trans. J. A. Smith, in The Complete Works of Aristotle, vol. 1, ed. Jonathan Barnes. Princeton. Princeton University Press.

Badiou, A. (2008) Number and Numbers, trans. Robin Mackay. Malden, MA. Polity.

Copjec, J. (1994) Read My Desire: Lacan Against the Historicists. Cambridge. MIT Press.

Hegel, G. W. F. (1991) Elements of the Philosophy of Right, trans. H. B. Nisbet, ed. Allen W. Wood. Cambridge. Cambridge University Press.

Lacan, J. (unpublished) Le Séminaire XXI: Les non-dupes errent, 1973–1974, unpublished manuscript, session of 11 December 1973.

McGowan, T. (2016) Capitalism and Desire: The Psychic Cost of Free Markets. New York. Columbia University Press.

Nabokov, V. (1962) Pale Fire. New York. Vintage.

Zizek, S. (1993) Tarrying With the Negative: Kant, Hegel and the Critique of Ideology. Durham. Duke University Press.

5

'To Be Homesick with No Place to Go': The Phantom of the *Sinthome* and the Joi of Sex

Daniel Bristow

Part I: The Phantom of the *Sinthome*

One way of articulating the *sinthome* could be by construing it as Lacan's (most) *artificial* concept. That is, it deals with—and deals *in*—*artifice*, much like the replica snakeskin, human-eyeball, or AI manufacturers that people market stalls, fabrication units, and the dark labyrinths of the Tyrell and Wallace Corporations in Ridley Scott's *Blade Runner* (1982–2007) and Denis Villeneuve's *Blade Runner 2049* (2017). Indeed, as Lacan stated of the *sinthome*: it ties things up—in the process (re-)engendering meaning—but it does so with *artifice*.[1] And yet—in Lacan, and *beyond*—it seems that the *sinthome* can come to *mean* almost anything—and it can—but this '*almost*' is very important: it is what keeps the *sinthome* in check (barring its *completion*—which would at the same time be its *reduction to zero*—of meaning).[2] It is thus representative of the equivocation that is deployable as *the only weapon against the*

D. Bristow (✉)
Philadelphia Association, Hampstead, London, UK

C. Neill (ed.), *Lacanian Perspectives on Blade Runner 2049*,
The Palgrave Lacan Series,
https://doi.org/10.1007/978-3-030-56754-5_5

sinthome, as Lacan enigmatically describes equivocation's role in interpretation in his twenty-third Seminar, to which the concept gives a title: *The Sinthome*. (Thus, in keeping an equivocality afloat, the avertability of totalisation, or of total loss—indeed, of *both at once*—of meaning becomes available and renewable, in analysis, and *beyond*.) Yet, it might be well to rephrase things slightly, and propose equivocation as *the other side* of the *sinthome*; that is, we may suggest that the two—equivocation/*sinthome*—should work in combination, or as a 'combinatory'. Indeed, there is already pronounced ambiguity in this area in the text(s) of Seminar XXIII, which oscillates between *sinthome* and symptom, terms which carry different (though not mutually exclusive) modulations that it will be worth pausing on, and disambiguating to an extent here first.

Rather 'in character', between the official French and English editions of the Seminar—and the 'pirate' translations into English—there is an equivocally curious distribution of terms. In the official French edition, edited by Jacques-Alain Miller and published by Seuil in 2005, and in Cormac Gallagher's translation, from unedited French manuscripts, it is—in accord with the above—the *sinthome* that is referred to at this point in the first session of the Seminar, of 18 November 1975; yet in the official English translation (of the same text as established by Miller) by Adrian Price, and in Luke Thurston's unofficial translation (of the *Ornicar?* journal serialisation of 1976–1977, also established by Miller), 'symptom' is used: '*en fin de compte nous n'avons que ça, l'équivoque, comme arme contre le sinthome*' (Seuil edition); 'when all is said and done, that is the only weapon we have against the *sinthome*: equivocation' (Gallagher); 'when all is said and done, the equivoque is all we have as a weapon against the symptom' (Price); 'in the end we only have that as a weapon against the symptom – equivocation' (Thurston).[3] Whereas the official English translation is closest to the original—if we can call it that, equivocation being eminently contagious in the process of Lacan's spoken words being stenographed 'live' ('*stécriture*', as he called this process)—in rendering the 'equivoque' as such, it departs from the *sinthome*, instead plumping for 'symptom'. However, as we stated above, we will not posit this as a mutually exclusive deadlock, but will here draw out the modulatory effects of each rendering. Indeed, as a weapon against the *sinthome*—or, as we have better phrased it: as *the other side* of the *sinthome*—equivocation can save *something of the subjective* from an engulfing totality, by fixing (a) meaning (this often in the shape of a symptom, as I have previously defined 'the other side of the *sinthome*', or—at the end of analysis—of an *assumed sinthome*). In the case of the symptom, equivocation can be used as a weapon against it as it can dislodge, deflect, or *de-range* (a certain) meaning, (a) meaning that's become *stuck*, which is so often what the symptom is, in its compromise formation as a *meaning-solution* (to a 'problem' insoluble by meaning alone; be[com]ing, in effect, a *half-said* solution): its giving a(nother) meaning to the traumatic kernel, even *navel*, it attempts—and fails, *wholly*—to repeat.[4]

Thus, whilst these brief ruminations should demonstrate how both uses are valid in their own ways in the equivocation equation, it is to the *sinthome* that we shall now return. With its late—almost crowning—conceptual appearance, the *sinthome* is liable to become (and has in places) something like a *hinge* of Lacanianism, a sort of knife-edge concept that can binarily be taken as that which either finally ties up Lacan's thought and teaching, or indeed as that which undoes it (even to the extent of supposedly exposing its faultlines and destabilising inconsistencies). Maintaining an equivocal place between these two ultimatums might rather be the best course of action, so as to not permanently close the subject of the *sinthome*, and in not doing so to also avoid foreclosing the sinthome*of the subject*.

In her magnum opus *What Is Sex?*—which will become a key text to the construction of this essay—Alenka Zupančič gives a very clear reading of precisely the above points regarding equivocation, relating them back to (her work on) the workings of jokes. Indeed, she states that 'in the case of jokes, playing on equivocity, and introducing another meaning with its help, does not have the result of relativizing meaning; it is, rather, designed so that we get a very precise, isolated point.'[5] Further,

> [an] equivocal punch line does not open up a multiplicity of possible meanings; rather, it evokes and uses this multiplicity in order to efficiently pin down and transmit a *singular point* (or impasse); and to transmit it

> in a most economical way—not by fully and exhaustively describing it, but by directly naming it: that is, precisely so that it functions somewhat like a formula.[6]

That which evokes and uses a multiplicity of meanings to transmit a *singular point* (i.e., a 'personality', or a memory) or impasse (mathemes of which, after Lacan, Zupančič strikingly later calls themselves 'the *formalization of the impasse of formalization*')—or indeed, again, *both at once*; e.g., the *divided* subject—thus fits almost precisely the formula for the *sinthome* of the '*empty signified*' that was originated for it in my earlier work on the relations between the writings and life of James Joyce and Lacan's late psychoanalytic theory.[7] We might suggest then that the *sinthome* is the *place* in which this multiplicity is animated, and is that which enacts the process of the selection and establishment of 'bits' of it as singular points of meaning. It is its interaction with equivocation—for a definition of which we might here annex a description of 'contradiction' in Zupančič: that 'which is *intrinsically* social—others, and our relation to them, as well as social relations more generally, are already implied in it'—that inaugurates this positionality (giving the shape of multiplicity to that which is within the *activated*—or *equivocable—sinthome*, as opposed to the formlessness of totality of the unactivated, or fallow, *unequivocal sinthome*; that is, the utterly *empty* signified).[8] To transpose this to the *Blade Runner* universe, it should be becoming obvious that what is at stake *for* replicants, or androids ('andys' as Philip K. Dick's Rick Deckard calls them), is what might constitute an appropriate response to the realisation that the selection and establishment from a multiplicity of meanings of a *singularity* (e.g., one's own subjectivity, one's sense of 'self') was not derived from a co-originatory interaction with the outside world (through human birth and upbringing; human-only attributes which of course get challenged in *Blade Runner 2049*), but was brought about by artificial implantation. To this reality Dick's mantra in *VALIS*, which is the epigraph to this part of the essay, will become its thesis question: is insanity the appropriate response, or is it a prescribed one?[9] An attempt to tease out an answer will be made by tracing the question's resonances through the *Blade Runner* films, films that in Scott Bukatman's words are 'deeply concerned with the making and unmaking of selves, and with worlds that are no longer *given*.'[10]

But first to return to the theme of resistance to totalisation and absolutism: we might offer here a twist on the modification of Bertrand Russell's set-theoretic paradox as it briefly appears, for example, in Jorge-Luis Borges 'The Library of Babel', in his reference to this library's (inevitable) inclusion of the fabled 'catalogue of catalogues'.[11] We will hit another aporetic wall if in a thought experiment we try to imagine *an encyclopædia of encyclopædias*; that is, a massive (multi-volume) tome proclaiming to contain in full all of the encyclopædias of the world, from first published to the latest. The fact that it itself is an encyclopædia would thus necessitate its inclusion of itself. If we imagine its contents pages listing the page numbers at which each encyclopædia begins, the last entry would therefore be for the very *Encyclopædia of Encyclopædias* itself, which would then have to reproduce itself *as such* (that is, as containing all the world's encyclopaedias, inclusive of itself) within its own pages, which naturally could not occur *exactly*, as a limit of self-inclusion is reached (page sequentiality would go askew, for example; i.e., continuity of page numbers would necessitate the reproduced *Encyclopædia of Encyclopædias*' contents pages being different to how they appeared at the start of the work; otherwise, its facsimile reproduction would negate page number continuity. Plus, of course, its re-inclusion of itself—as the final encyclopaedia in the set, each time round—would become a *mise en abyme* and *reductio ad absurdum*). Thus, it could only possibly contain itself *as other to itself* (much like a matryoshka doll, inside of which is not itself, but *an other of itself*).

In its constantly being thwarted in its attempt to (re-)include itself *as itself* we could see this *Encyclopaedia* as a manifestation of a(n impossible) desire to complete a (perfect) circle. Yet, there is always a gap, or a cut, a division, break, or divide (this theme is of course fundamental to Lacan's teaching at so many points): as *the place of reinscription; recirculation*. As such, we could depict this desire thusly (Fig. 5.1).

Whilst Lacan was most acutely aware of the *not-whole* status of his theoretical edifice—whilst its very constitutivity rests on this—the *sinthome* nonetheless could be deployed here to tie up this gap *with its*

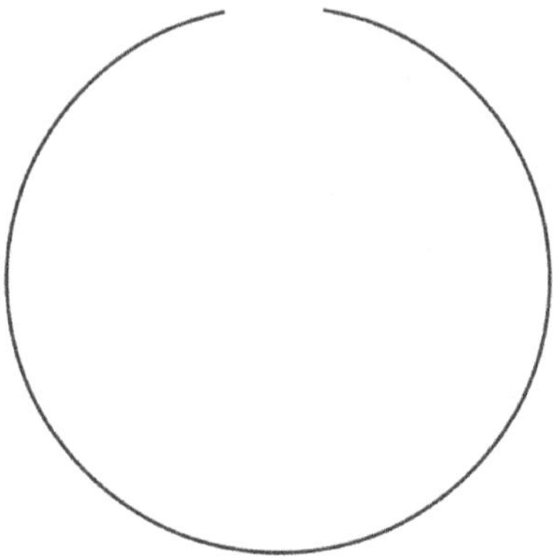

Fig. 5.1 Incomplete/incompletable circle of the desire for completion

artifice. Resultantly, if we subjectively transpose this, in the (contradiction produced by the) incompletion of knowledge, an analytic end can thus be found in coming to 'assume our *sinthome*', in the sense of it as our *irreducible symptom* (the evisceration of which would irrecoverably desubjectivise us). As Zupančič is at pains to remind us, however, psychoanalytically this *assumption* 'is not about accepting the contradiction, but about *taking one's place in it*.[…] This is what analysis ideally leads to: contradiction does not simply disappear, but the way it functions in the discourse structuring our reality changes radically. And this happens as a result of our fully *engaging in the contradiction*, taking our place in it.'[12] Thus, the *sinthome* is one way through which to radically change reality, but it is so strictly on the side of the individual (as Lacan said: 'the *sinthome* […] is *what is singular to each individual*').[13] That is to say: *it drops inside itself* (which is open to the rephrasing: *it drops inside its desire*) (Fig. 5.2).

If we figurate the incomplete circle as the *contradiction*—the very incompletion, the *navel*, of which is its structurality—the (assumption of the) *sinthome* becomes a means by which to 'take our place in it'. This is a radical analytic point the centrality of which must be thoroughly stressed. What this *assumption* of the *sinthome* can do is produce 'a new signifier', as Lacan once said, which will have modulatory effects on subjectivity. It can be produced, as Zupančič states, quoting Lacan,

Fig. 5.2 Assumption of sinthome 'tying up' the gap

'"starting from the efflorescence of the signifier," its polysemic babble, its equivocities' (that is, those which arise in analysis: the *talking cure*).[14] And by this new signifier a 'new *subjectivation* [can be] triggered'; put another way, by Zupančič: 'only a *letter* can disentangle what exists only in entangled form, and hence eventually change this form itself.'[15]

However, this essay will come to argue that there might also be another emancipatory way (perhaps *after* the assumption of the *sinthome*, in general subjective terms), one which could even be necessitous to the type of subject that might find themselves in a position analogous to certain of the replicants in the *Blade Runners*. For it is this position that gives a parallax view of the *sinthome*, and might show that there is what we could call a '*phantomic*' element to it (just as there is a *sinthomic* element to all phantasy, or phantasm): that it is, due to its *artifice*—however usefully deployable this might be—something of a *phantom* in itself.

Thus, for replicants, the *sinthome* is, or can become (to play with the subtitle of the first *Star Wars* episode) a *phantom menace*; that is, in general, its hardware being liable to faults, etc. (there would have likely been throughout replicant evolution models prior to the Nexus series, and especially to the Nexus 6, that would have been hampered by defects in their hardwiring, shortcircuiting subjectivity, for example), but especially when a replicant itself is awoken to consciousness of it, consciousness of the status of one's ownmost memories in fact being 'implants [...] somebody else's', as Deckard (Harrison Ford) tells Rachael (Sean Young) hers are in *Blade Runner*, and as K (Ryan Gosling) has found out his are in the sequel.[16] What thus *sinthomically* separates the human and the replicant subject here is that the human is *by its 'nature'* foreclosed from the knowledge of its *sinthome*, or if it has analytically *assumed* this, it can—perhaps *must*—nonetheless disavow it. A replicant, when aware of its *sinthomic* implantation, is foreclosed from the possibility of this disavowal (to attempt the disavowal might be a fruitless endeavour, whereas for the human subject to *insist* on the *sinthome*—its implantation; hypothesised in theories of design and so on—might be its).[17] Thus, for replicants, it can be argued here that it would be false (consciousness)—or rather *symptomal* or symptomatic—to accept this fate (simple knowledge of *sinthomic* implantation) *as* (one's) *sinthome* and assume it. As is quite clearly the case with K, this leads to an *insanity of conformism* (the agreement to avoid *malfunction*; that is, any cultivation of *subjective* affects such as empathy—tested for in the intense baseline assessments, and suggested by the suspicion and derision the notion of a 'soul' in androids is met with, redolent of its status also in the ideological conjuncture of Yevgeny Zamyatin's *We* (1921)—although a regulated *desire* [of the Other] is allowed), and this in opposition to the (suicidal) *insanity of exile,* which the replicants of the second film are given the forced choice between (a forced choice much in the vein of Judge Scarecrow's [Cillian Murphy] in *The Dark Knight Rises* [2012] between death or exile, which, when the former is chosen, engenders the sentencing: 'Very well. Death... by exile'). In the case of K's *administered desire*—that for the pleasure application of Joi (Ana de Armas) (whose own subjective status is central to, and becomes a pressing issue in, the movie, as will be explored in the second part of this essay)—we see the implementation and intensification of what Alfie Bown describes in his *Playstation Dreamworld*: that 'the principal function of the virtual technological space is not to let us be who we want to be, nor to give us what we want, but to change what and how we desire'; here to bring it in line, set it to the patterns of conformity.[18]

Prior to *reconfiguration* to either of these insanities, the replicant subject thus becomes radically emptied out, in the face of (their—new, more *real* –) *reality*. What has become exempted, or subtracted, from the *sinthome* in this *real*isation is precisely its *-home* element; that is, there is nowhere—or, there is no longer anywhere—*where the heart is*.[19] As the replicant Leon Kowalski (Brion James) puts it in an early version of Hampton Fancher and David Peoples' script for the first film (in a scene which was filmed, but deleted): this is precisely 'to be homesick with no place to go.' Yet the trajectory that K has ended up embodying (one hinted at again in the first version of the *Blade Runner* script in the line: 'pretty soon the public will want skin jobs for Enforcement')—in the early parts of the film, at least—is one that over the course of the years that separate the movies and their events has only moved from *slavery to slavery*: from the slavery in the Off-world colonies, for which labour the replicants were originally engineered, to the (albeit paid) slavery to the state apparatus—and subscription to its ideology—in blade-running and 'retiring' other replicants, in full knowledge that one is one oneself, a form of servitude that speaks to long traditions of oppressed being delegated the very work of oppressors, by them, and their state apparatuses). It is thus within the (*sint*) *homelessness* of this contradiction that a place should perhaps initially be taken, as it may be from this place that another world—another *sanity*—beyond it can be envisioned.

Part II: The Joi of Sex

> Sex, reproduction, security, the simple things. But no way to satisfy them. To be homesick with no place to go. Potential with no way to use it. Lots of little oversights in the Nexus Six.
>
> —Leon, *Blade Runner* screenplay, Hampton Fancher and David Peoples.

Whilst this specific homelessness is rife amongst the replicants of the *Blade Runner* universe, namelessness is not, and it is through the *name* that (the signifier's) *determination* comes to be produced. In Lacan's words concerning witticisms in Seminar V, the name, too, can 'designate, and always off to the side, what is only ever seen by looking elsewhere.'[20]

We will take three examples here from the *Blade Runner* films: those of K, Joi, and Rachael. K (full identification code: Officer KD6-3.7) is a character in effect without a (real) name, as Joi points out to him, subsequently *renaming* him 'Joe'. This nominative transformation is extremely suggestive; in fact, suggestive of how this renaming is not actually a transformation at all, but a tripartite prolongation. Firstly, the name *exposes* Joi's own: to make 'Joi' all that is needed is the addition of the lowercase '*i*'—the letter so ubiquitous in today's *self*-centred personal technology (Apple's iPod, iPhone, iPad, and, most suggestively, *i*(Pod)*Touch*; the BBC's iPlayer, etc.)—to 'Jo' (or, rather, the replacement of the 'e' for 'i' in the male name).[21] Thus, in this is exposed her very (at least corporately intended) role as only *supplement* or *surrogate* to K (whether or not he is K or Joe). (Fuller exploration of Joi's, and of *the woman's*, subjectivity will take place in the remainder of this essay.) Secondly, the new name is anything but 'special'—as Joi intends it to be, to nominate her *special* partner—in that it is the very name that the genus of her particular pleasure model (the Wallace Corporation's 'DiJi' digital companion) uses to designate and attract—that is, precisely to *interpellate*—its male audience (as demonstrated by the supersized interactive holographic advertisement for DiJis that hits on K as he walks past). Thirdly and finally, the journey from K to Joe in fact travels no distance, if relayed against the most famous, and no doubt influential, of the names in Franz Kafka's work: K of *The Castle*, and Josef K of *The Trial*. Thus, K/Joe becomes just another subject of the bureaucratic and faceless apparatus of a space that much resembles those of Kafka's inscriptions.

Joi's name is determinative in its very ancillariness and supplementarity. In the most diabolically *gendered* way, her engineering—her very *engendering*—makes transparent her role as *woman* created in the image of *man's symptom*. Thus, this is the subjectivity that Joi is ingeminated into, and her name—itself an acronym in a pornography subcategory for 'Jerk-off instructions'—dictates the limits of the place she is able to take within it, and the borders of the space that she is able to inhabit. Yet, Rachael is possibly the most determined of these names, certainly classically. The name itself means 'ewe' in Hebrew, making her the very 'electric sheep' of the title of Dick's original novel on which *Blade Runner* was based: *Do Androids Dream of Electric Sheep?*[22] Further, biblically, even her very fate in the sequel—her remains forensically indicating that she had died from complications that arose in carrying out a caesarean section—is also that of the biblical Rachel, who died in childbirth.[23]

Thus, whilst names are symptomatically predeterminative to a great extent in the *Blade Runner* movies—laced with fatal inscriptions—replicant agency is eminently open to something like *sinthomic* creativity. In his analysis, Bukatman emphasises the *performance* that Roy Batty (Rutger Hauer) puts on in his dying stages when he has gone from being the hunted to being the hunter, chasing Deckard (*literally*) *through* the Bradbury Building (clambering around inside its walls and recesses, and bursting out of them through brick and tile, scaling its structural girders, forcing Deckard to climb through its ceilings/floors), all the while displaying a particularly camp machismo, often 'feigning to feign' (by 'mak[ing] tracks whose deceptiveness lies in getting them to be taken as false, when in fact they are true—that is, tracks that indicate the right trail'—Lacan's criterion for what distinguishes the human from the animal (the *replicant* throwing a spanner in the works of this dichotomy)—scrambling Deckard's lines of flight), yet also becoming-animal, in his repeated wolflike howls.[24] Earlier in the film, Pris (Daryl Hannah) performs pirouettes and cartwheels, her body being utilised to its full potential (being *lived through*, in its state emancipated from the torpor of uncertainty, the replicants' deaths—as they are all too aware—fast approaching, either 'naturally', by the expiration of their four-year lifespans, which they now know they cannot alter, or enforced, by 'retirement' by a blade runner, such as were Zhora (Joanna Cassidy) and Leon's fates). It would appear that the consciousness of their replicant state—and this state's fate—enables the replicants of *Blade Runner* to mould their mode of being to their own desire, and to explore the very limits of its performativity (that is, of what is performable within the modality of being that they have been imbued with). Yet, due to the constriction of the individual-yet-common fate with which they have collectively been endowed (their four-year lifespans and thus certain deaths at a predetermined time), this performativity is in itself constrained.

The outlet that performance finds in the sequel is however of a different stamp. If Judith Butler in the 1999 preface to her landmark work *Gender Trouble* (1990) defines 'performativity [a]s not a singular act, but a repetition and a ritual, which achieves its effects through its naturalization in the context of a body, understood, in part, as a culturally sustained temporal duration', then in *Blade Runner 2049 sex* seems to have been performatively overcome, subversively deterritorialised, whilst *gender* (partaking of what Lacan tabulates in his logics of sexuation) has absolutely not—its stereotypical roles and encoded performativities being disastrously entrenched—whilst *sexuality* and its associated phobias have been *displaced* (onto prejudices concerning inter-tech sexual desires, for example).[25] In effect, gender has become dystopian (although, that is, only as an elongated version of today's gender dystopia), whilst *dysphoria* has become *general*, *out-of-placeness* being the very mode of being in 2049 on Earth, dictated by environmental factors and subjective status.

In Tanya Lapointe's book on the making of the sequel, *The Art and Soul of* Blade Runner 2049, the film's executive producer Bill Carraro states: 'in the future, sex and sexuality aren't taboo. So it's natural to go into a bar and, before getting a hamburger, hav[e] some sex.'[26] As Lapointe explains—and as is opaquely depicted in the movie— 'a bar named Bibi's [is] where people have sex just as casually as they would a conversation[.] Villeneuve ma[kes] it clear that Bibi's Bar is not a brothel. The purpose of the establishment is to offer a safe and playful environment to have sex with friends, strangers, or significant others[,] open to female and male customers alike.'[27] Whilst it is too simplistic to suggest that sex and sexuality *in toto* aren't taboo in the world of the sequel—the reconnoitring replicant prostitute Mariette (Mackenzie Davis) is initially discriminatory against DiJis, for example (she says to K: 'oh, I see, you don't like *real* girls')—there has nonetheless been a seismic shift, which has led, for example, to transitionalities in embodiment, embodied performativity, costuming and sartorial signification becoming much more accepted and widespread. Indeed, after her initial qualms (as seductively performed as genuinely meant, perhaps), Mariette—in whose name resonates the signifiers 'marionette', even 'maquette'—*becomes* Joi's body for K to experience sexually: they meld by inhabiting the same space at the same time, one—Mariette—physically, over whom the other—Joi—is able to superimpose herself, holographically. Whilst this quite exceptional threeway trans experience breaks through many 'taboos' that delineate the sidelines and margins of today's sexual mores, gender inscriptions and delineations within the Los Angeles megacity are grossly dichotomous, between male and female, and their symbolic functions; woman, as ever, is objectified, precisely in the psychoanalytic sense, of becoming the *object* of man('s refinding), as Lacan describes with such clarity in Seminar IV:

> From the moment of her encounter by man, the feminine subject is always called upon to inscribe herself in a sort of refinding, which places her from the start in a position characterized by the ambiguity between natural relations and symbolic relations. The ambiguity is precisely where, as I am trying to demonstrate, the analytic dimension resides.[28]

Yet, whilst being the very place of residence of the analytic dimension, this dimension itself is one that gets avariciously subsumed into—and exploited by—capitalist discourse. In the capitalist metropolis of 2049, *choice* is still the major engine of perpetuating motion, and reliance on the commodification of the female and the feminine still the major trope that *sells* choice (choice itself, and its endless renewal ('refinding'), is in fact *the* commodity, the *Ur-commodity*). What is (at least) fantasised in this is that *the man* is the decider, the connoisseur, chooser, and *woman* is not only always one *of* many—one of an array to choose *from*—but also one *as* many (a DiJi can morph into whatever the man desires her to be at that moment, as Joi showcases when we first meet her). The masculinist desire (masculinised, discursively; through cultural reinscription, etc.) that underwrites this aspect of capitalism reveals again said desire's own thwarting (and thus further reinscription) in that it goes to expose what Freud's 'myth of the enjoyment of *all the women* [in *Totem and Taboo*] designates[;] that there are not *all the women.* There is no universal of the woman', as Lacan states in Seminar XVIII.[29] Yet, the mytheme of '*all the women*' is one of the key signifiers to the patriarchal economy's functioning, an economy that plays the easiest (according to the historicity of today's capitalist conjuncture; its *hitherto*) into the hands—or 'handjobs', to evoke Zupančič's rewording of Adam Smith—of the market. As Jacques Derrida pertinaciously and historiographically uncovers in *Spurs*, the very seeds of the gendered capitalist *epistēmē* are buried deep within the ontology that constitutes its foundation: 'although there is no truth in itself of the sexual difference in itself, of either man or woman in itself, all of ontology nonetheless, with its inspection, appropriation, identifications and verification of identity, has resulted in concealing, even as it presupposes it, this undecidability.'[30]

Woman in *Blade Runner 2049* thus functions as man's lost object, his symptom.[31] This construction of *woman*—'*The Woman*' *qua* economical signifier—thus establishes 'her' (that is, her male-imposed, and -dictated, subjectivity) as inexistent—Lacan of course crosses through the feminine article '*la*' to designate this in Seminar XX—beyond which lies the—necessitous, due to patriarchal economy's exclusionariness—autopoiesis of *écriture feminine*, and the very necessity of feminism and feminisms. In a similar respect, it could be said that '*le Sinthome n'existe pas*' ('*the Sinthome* does not exist') for the replicant subject: it is—as we have said before, and as applies to (the notion of) 'The Woman'—*phantomic*. Replicant subjectivity cannot '*drop inside itself*'—in the sense we formulated for the *sinthome* above (in effect it is 'hyper-Möbial' in that its inside *is its outside*, and that this is *known*); it has no place to go, *within*: it is *sinthomesick*.

Thus, what option is left open to it is glimpsed in the second film in the form of the replicant resistance, which is geared towards a form of futural utopia (if only for replicants and not the totality of life on Earth). The plotline is rather undeveloped and vague in *Blade Runner 2049*, but it draws on the history of resistance in and between the two movies (for example, action taken by replicant activists in 2022—depicted in the short film *Black Out 2022* (2017, directed by Shinichirô Watanabe)—leads to a global powerout, the closure of the Tyrell Corporation—as it was Nexus 8 replicants who orchestrated the events—and the illegalisation of replicant production, as well as achieving its intention of deleting the digital data files held on the replicant population). The collectivisation of the replicants in this movement—headed by the rebel leader Freysa (Hiam Abbass) (an Oedipal—or Tiresian—replicant, symbolically shown by the absence of her right identification-code-watermarked eye), and of whom Mariette is a part, and who sequester K—hints at the cut that can circumvent the circumscriptions of the (replicant) *sinthome*, redistribute its proprietorial ownership, and bring about its reclamation:

stretch it, remould it, from the individual to the communal. That is to say—or to put forward the hypothesis that—*the utopian* might *draw it, the* sinthome *(and the* jouissance *that is its lifeblood), out of itself* (out of *its* desire into a *communal* desire). Thus, the communal utopia becomes an other side of the individual *sinthome*, in being a form 'more redolent of perversion than paranoia, and with that passionate sense of mission or calling from which *jouissance* is never absent', as Fredric Jameson defines the utopian at one point in *Archaeologies of the Future*.[32] The redirection of the flow of *jouissance*—experienceable (to say nothing here of how *fully so*) by *experiential forms*, replicant and human alike—constitutes a drive or *pulsion* counter to the *sinthome* (and the enforcement of its assumption in service to an oppressive state), if diagrammatised according to the above-used terms (as in Fig. 5.3).

What can thus be envisaged here is that through the utopic might be unlocked just as much the home of a subjectivity with no place to go as the joy of a sex which is not one. Thus, for the *Blade Runner* movies, in their science fictional aspect, the discontents of the phantom of the *sinthome* in replicant subjectivity are on a par with those of sexual difference (represented in the socio-cultural and politically factual

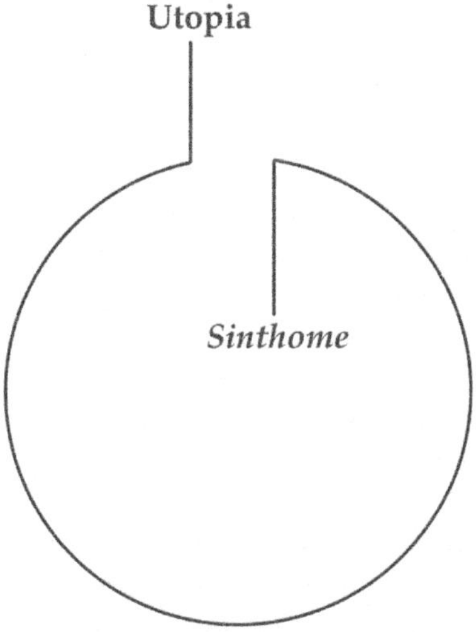

Fig. 5.3 The utopian as non-sinthomic recathexis

aspects of the films), which Luce Irigaray famously described as 'probably the issue in our time which could be our 'salvation' if we thought it through.'[33] *2049* stresses the need for this thinking-through; for utopia to be envisioned; and for the *sinthome* not to stop being written anew.

Notes

1. See Lacan (2016a; 58): 'finding a meaning entails knowing which knot it is, and joining it together fast by means of an artifice.'
2. To render the coincidence of *completion* and *evisceration*—or *reduction to zero*—*at once*, we might say that these are not two sides of the same coin, but rather, in *coinciding*, their *total coincidence* is that which obliterates the coin itself, for it is the coin—acting in the commonplace phrase in the capacity of a *screen* (that which resists the totality before or beyond it)—that *gives* its sides, through the process of *separating* and thereby *uniting* them. This process is elaborated in much more detail in my reflections on *enverity* in Bristow (2016). In Jacques Derrida's work, like resistance of absolutism is echoed in the dual conceptual strands of his oft-repeated parameters 'almost-everything' and 'almost-nothing', which in themselves—to stretch a phrase of the philosopher's—give 'the reasons why a totalization is impossible' (see Derrida [2013; 136]). Such is due to the '*almost*' keeping totalisation ('everything'/'nothing') at bay and preserving the *possibility of an infinity*, by pushing totalisation to an infinite distance ('*almost-everything*'), or enabling infinite regress ('*almost-nothing*'; that is, a nothing which can never be quite got to).
3. See Lacan (2005; 17) [my italicisation], Lacan (Unofficial translation, n.d.-a; session I, 10), Lacan (2016a; 9), and Lacan (?; 4).
4. This sliding scale between *sinthome* and symptom is somewhat similarly represented in that depicted between '*interdit*' ('dumbfoundedness') and '*souffle*' ('flabbergastation')—between which ends a hold on meaning can be kept, to varying degrees in Bristow (2016; 172, ff).
5. Zupančič (2017; 65).
6. Ibid.; 66.
7. Ibid.; 69. See Bristow (2016). 'Directly naming' the *singular point* also speaks precisely to the *sinthome*, in its relation to nomination, and specifically Joyce's *making* a name for himself.
8. Zupančič (2017; 66).

9. See Dick (2001; 10).
10. Bukatman (2012; 19).
11. See Borges (2000; 78–79).
12. Zupančič (2017; 72).
13. Lacan (2016b; 147) [my italics].
14. Zupančič (2017; 127).
15. Ibid.; 128.
16. Thus, to describe the replicant *unconscious* that this form of *sinthomic* subjectivity produces, Philip K. Dick's aporic description, in the novel on which the film is based, of 'a rough, cold android, hoping to undergo an experience from which, due to a deliberately built-in defect, it remained excluded' provides a perfect summation. See Dick (2017; 146).
17. It will be worth reading these reflections against those made by Slavoj Žižek in the subsection '*"Total Recall": Knowledge in the Real*', in Žižek (1993; 39 ff.). Indeed, to force the notion of the total separation of replicant and human subjectivity would also be to dissolve the patiently articulated anti-humanism that Lacan based so much of his work on. Slippage, in the *ideation*, between human and android, is demonstrated pertinaciously by Dick when a human imagines a bounty hunter (of which he has never before heard) *as precisely the machine it is on its mission to annihilate*: 'he had an indistinct, glimpsed darkly impression: of something merciless that carried a printed list and a gun, that moved machine-like through the flat, bureaucratic job of killing. A thing without emotions, or even, a face; a thing that if killed got replaced immediately by another resembling it. And so on, until everyone real and alive had been shot.' See Dick (2017; 125).
18. Bown (2017; 91).
19. In *Do Androids Dream of Electric Sheep?*, Dick describes this confrontation—as made by the bounty hunter Phil Resch, whose ontological status is thrown into question—in relation to Edvard Munch's painting, *The Scream*: 'the painting showed a hairless, oppressed creature with a head like an inverted pear, its hands clapped in horror to its ears, its mouth open in a vast, soundless scream. Twisted ripples of the creature's torment, echoes of its cry, flooded out into the air surrounding it; the man or woman, whichever it was, had become contained by its own howl. It had covered its ears against its own sound. The creature stood on a bridge and no one else was present; the creature screamed in isolation. Cut off by—or despite—its outcry.

"He did a woodcut of this," Rick said, reading the card tacked below the paining.

"I think," Phil Resch said, "that this is how an andy must feel." He traced in the air the convolutions, visible in the picture, of the creature's cry. "I don't feel like that, so maybe I'm not an —" He broke off, as several persons strolled up to inspect the picture.' See Dick (2017; 104). On Lacan's interaction with *The Scream* in Seminar XII—which he describes in similar terms to Dick—in relation to another classic sci-fi film, see Bristow (2017).

20. Lacan (2017; 19).
21. In effect, this inescapability of the *i*-society slightly agitates Denis Villeneuve's original prescription for the construction of the 2049 universe in which the sequel is set, as outlined in his revealingly capitalistic, albeit poetic, comment: 'I put together a team of dreamers, to create the future of an old dream. I asked them to imagine the future without Steve Jobs, but with the USSR.' See Villeneuve (2017; 7).
22. Although the final film is only loosely based on the plot of Dick's novel, its title has a looser still connection to its origin, a movie treatment that William S. Burroughs wrote in adaptation of a novel by Alan E. Nourse (*The Bladerunner*, 1974). See Burroughs (1999). Fancher suggested the name from a copy of Burroughs' treatment that he had, in preference to the working titles *Android* and *Dangerous Days*. For more on this evolution, see *Dangerous Days: Making* Blade Runner (2007, dir. by Charles de Lauzirika).
23. See Gen. xxxv. 18. The plotline involving Rachael's imperfect (in effect, matricidal) birth-giving is one prophesied more optimistically by a much earlier B-movie, as Bukatman explains in his companion to *Blade Runner*: 'in *The Creation of the Humanoids* (1962), a terrific low-budget precursor to *Blade Runner* written and directed by Wesley Barry, humanoids are becoming disturbingly perfect simulacra. The protagonist is a member of the reactionary Order of the Flesh and Blood, but he turns out to be a humanoid after all, and the other humanoids give him an operation to upgrade him to the new, sexually reproducing, model R100. "Of course, the operation was a success," the narrator assures us at the end, "or *you* wouldn't be here."' See Bukatman (2012; 76).
24. See Lacan (2006; 683).
25. Butler (2006; xv).
26. Bill Carraro, quoted in Lapointe (2017; 68).
27. Lapointe, in ibid.

28. Lacan (Unofficial translation, n.d.-b; 101).
29. Lacan (Unofficial translation, n.d.-c; session IV, 16).
30. Derrida, J. (1979; 103–105).
31. The characters of Lieutenant Joshi (Robin Wright) and Luv (Sylvia Hoeks), for example, of course offset this, but are also quite an exception to the systematicity on which the 2049 megacity and its culture operates.
32. Jameson (2007; 10).
33. Irigaray (1993; 5).

References

Borges, J.-L. (2000) 'The Library of Babel', trans. by James E. Irby. *Labyrinths: Selected Stories and Other Writings* [1964], ed. by Donald A. Yates and James E. Irby. London. Penguin Modern Classics.

Bown, A. (2017) *The Playstation Dreamworld*. Cambridge. Polity.

Bristow, D. (2016) *Joyce and Lacan: Reading, Writing, and Psychoanalysis*. Abingdon. Routledge.

Bristow, D. (2017) *2001: A Space Odyssey and Lacanian Psychoanalytic Theory*. Cham. Palgrave Macmillan.

Bukatman, S. (2012) *Blade Runner* [1997]. London. Palgrave Macmillan/British Film Institute. 2 edn.

Burroughs, W. S. (1999) *Blade Runner: A Movie* [1979]. Berkeley. Blue Wind Press.

Butler, J. (2006). 'Preface (1999)'. *Gender Trouble: Feminism and the Subversion of Identity* [1990]. Abingdon. Routledge Classics.

Derrida, J. (1979) *Spurs: Nietzsche's Styles/Éperons: Les Styles de Nietzsche* [1978], trans. by Barbara Harlow. Chicago and London: The University of Chicago Press.

Derrida, J. (2013) 'Like the Sound of the Sea Deep within a Shell: Paul de Man's War' [1988], trans. by Peggy Kamuf, in *Signature Derrida*. Chicago. University of Chicago Press.

Dick, P. K. (2001) *VALIS* [1981]. London. Gollancz.

Dick, P. K. (2017) *Do Androids Dream of Electric Sheep? [Filmed as* Blade Runner*]* [1968]. London. Gollancz.

Irigaray, L. (1993) *An Ethics of Sexual Difference* [1984], trans. by Carolyn Burke and Gillian C. Gill. Ithaca and New York. Cornell University Press.

Jameson, F. (2007) *Archaeologies of the Future: The Desire Called Utopian and Other Science Fictions* [2005]. London. Verso.

Lacan, J. (2005) *Le seminaire, Livre XXIII: Le sinthome* [1975–1976], texte établi par Jacques-Alain Miller. Paris. Éditions du Seuil.

Lacan, J. (2006) 'The Subversion of the Subject and the Dialectic of Desire in the Freudian Unconscious' [1960]. *Écrits: The First Complete Edition in English* [1966], trans. by Bruce Fink, in collaboration with Héloïse Fink and Russell Grigg. New York. W. W. Norton.

Lacan, J. (2016a) *The Seminar of Jacques Lacan, Book XXIII: The Sinthome*, ed. by Jacques-Alain Miller, trans. by Adrian Price. Cambridge. Polity.

Lacan, J. (2016b) 'Joyce the Symptom'. *The Seminar of Jacques Lacan, Book XXIII: The Sinthome*, ed. by Jacques-Alain Miller, trans. by Adrian Price. Cambridge. Polity.

Lacan, J. (2017) *The Seminar of Jacques Lacan, Book V: Formations of the Unconscious*, ed. by Jacques-Alain Miller, trans. by Russell Grigg. Cambridge. Polity.

Lacan, J. (Unofficial translation, n.d.-a) *The Seminar of Jacques Lacan, Book XXIII: Joyce and the Sinthome, 1975–1976*, trans. by Cormac Gallagher. London. Karnac.

Lacan, J. (Unofficial translation, n.d.-b) *The Seminar of Jacques Lacan, Book IV: The Object Relation*. ed. by Jacques-Alain Miller, trans. by L. V. A. Roche.

Lacan, J. (Unofficial translation, n.d.-c) *The Seminar of Jacques Lacan, Book XVIII: On a Discourse That Might Not Be a Semblance, 1971*, trans. by Cormac Gallagher. London: Karnac.

Lacan, J. (Unofficial translation, n.d.-d) 'Le Sinthome: Jacques Lacan's Seminar XXIII, 1975–6', ed. by Jacques-Alain Miller, *Ornicar?* 6–11 (1975–1976), trans. by Luke Thurston.

Lapointe, T. (2017) *The Art and Soul of* Blade Runner 2049. Los Angeles. Alcon Entertainment.

Villeneuve, D. (2017) 'Foreword'. Lapointe, T. *The Art and Soul of* Blade Runner 2049. Los Angeles. Alcon Entertainment.

Žižek, S. (1993) *Tarrying with the Negative: Kant, Hegel, and the Critique of Ideology*. Durham. Duke University Press.

Zupančič, A. (2017) *What Is Sex?* Cambridge, MA. MIT Press.

6

Home Bodies: Prosthetic People and Economies of Desire

Timothy Richardson

Holograms and statues of women, variously undressed, crowd much of the frame of *Blade Runner 2049*. In deserted Las Vegas, enormous naked women are on their knees. In Los Angeles, huge ballerinas pirouette and pink and purple nudes bend low to flirt. In this world, vulgar fantasy has been built out to the overtness of ideology. It calls to mind Slavoj Žižek's discussion—very early in *Plague of Fantasies*—of Soviet statuary on top of office buildings where "the tendency to flatten the office building (the actual workplace for living people) more and more became clearly discernible, so that it changed increasingly into a mere pedestal for the larger-than-life statue… an ideological monster which crushes actual living men under his feet" (1–2). Žižek's point is that ideology is on the surface. It is obvious. In *Blade Runner 2049*, ideology is fantasy writ large.

T. Richardson (✉)
Department of English, University of Texas at Arlington, Arlington, TX, USA
e-mail: trichardson@uta.edu

C. Neill (ed.), *Lacanian Perspectives on Blade Runner 2049*,
The Palgrave Lacan Series,
https://doi.org/10.1007/978-3-030-56754-5_6

At this point, we should probably take as axiomatic Lacan's claim that sexual rapport does not exist, and one of the jobs of fantasy is to cover over that impossibility. Just as, per Žižek in the same book, "'ideology' is the 'self-evident' surface structure whose function is to conceal the underlying 'unbalanced', 'uncanny' structure," fantasy functions as an elaborate scaffolding to both misdirect us from and to cover over the fundamentally missing relationship (106). And Lacan asserts that masculine fantasy in particular depends on objectification in a full sense: "It is inasmuch as [object cause of desire] plays the role somewhere—from a point of departure, a single one, the male one—of that which takes the place of the missing partner, that what we are also used to seeing emerge in the place of the real, namely, fantasy, is constituted" (63). Frequently, these fantasies and their objects take standard forms and received traditions that are distilled for private enjoyment. The giant women of *2049* are advertisements for products and services that may have once been illicit, but that have since become publicly profitable. Not only are they ubiquitous, some are so massive that they can only be enjoyed from a distance. They are like architecture that has become commonplace and boring. The whole world is a sad Las Vegas. And as in Las Vegas, fantasy as ideology welcomes capital.

The commodification of women isn't new, of course, but *Blade Runner 2049* amplifies the practice to near universality. There are, after all, few "real" women of consequence in the film. Instead, the female characters stand for ideas or ideals for the men and, as the film emphasizes, these are old ideals of femininity (the prostitute and housewife) that are marketed and purchasable. Additionally, the masculinity the film offers is by-and-large white and bourgeois. But it is probably important to note that there are also few "real" men in *Blade Runner 2049*.

By almost every account, *Blade Runner 2049* is a beautiful film that, by some of the same accounts, has a real problem with women.

Certainly, the first *Blade Runner* (1982) does, too. There is not one human woman in it. Of the three female characters, each is a replicant and all but one is "retired." The surviving, Rachel, is loved. The original film takes as its broad theme a definition of humanity and, because it must, highlights the masculine and capitalistic supports for that act of defining.

The struggles of the film are always between men: Rick Deckard's manhandling by the regal and tragic Roy Batty, the father-son tension between Batty and Eldon Tyrell, and even Tyrell's patronizing chess games with J. F. Sebastian. Female characters are support for the various narratives and are not, by both definition and filmic treatment, alive. Of course, some of the men are not alive, either.

Because they are a workforce manufactured by the Tyrell Corporation, replicants have a limited lifespan by design. Some resist their lot and need to be handled. The policing we see in the film is devoted to trimming these excesses, civil service thus portrayed primarily as the protection of the interests of the more-than-global Tyrell Corporation, headed by its myopic god. Business—replacing governments of older, quainter dystopias—is the only activity in a film that has since become the standard of cyberpunk dystopias. Corporate enterprise is less an efficient or material cause for action, but some sort of collision of formal and final causes. It regulates every aspect of human living—its representatives hover in the skies, looking down—and has become the origin of something approaching human life. Life, we are shown, happens under colossal advertising, as Big Brother is now looking to exploit markets more that citizens.

Blade Runner 2049 both extends and reduces the positions of its predecessor. To the degree that (almost) every woman isn't human, the sequel echoes the first film. Though *2049* gives us many more female characters, all but the police chief and the dream architect are what Kenneth Burke calls a *purchasable miracle*, "loved not merely 'for herself,' but for what she 'represents,' as charismatic vessel of a social motive" (278). In the first film, Rachel is a body of memories of child-like lost beauty whom the hero saves because, well, he loves her, and Pris is a "basic pleasure model" for the use of soldiers (and, also, an accomplished acrobat). In the later film, Joi is one of many purchasable holographic companions who can change her appearance to suit an occasion and an owner's preference and who—at least in her private, domestic version—is as tied to the home as any traditional wife (unless her owner purchases for her a new peripheral gadget to take her on the road; this is like buying her shoes). Joi is a much more advanced, if more accommodating and ephemeral, pleasure model. And even the dream architect is ultimately revealed to stand for a fundamentally social motive. The police chief dies, since she has no place in the market.

Given this gloss, defending the presentation of women in either *Blade Runner* film would be difficult. I won't even try. The films have, on the one hand, very little useful to say about the feminine as such. The versions of femininity they propagate service a received, traditional masculinity and work to cover over the fundamental impasse that is called sexual by Lacan precisely because it is impossible except as a fantasy. On the other hand, *Blade Runner 2049*—even more than its predecessor—has a great deal to offer as it models the impasse of (vulgar, received, masculine) fantasies of mastery and their translation from private making-do to a public bazaar. Fantasy, it argues, has a problem with women. Of course, that problem with women indicates a much more substantial, foundational problem, as, per Alenka Zupančič, "The lack of sexual relation is real in the sense that, as lack or negativity, it is built into what is there, determining its logic and structure in an important way" (*What is Sex* 18). Fantasy engages desire in such a way as to deal with the problem of fundamental lack of which the "women" of the films are one symptom. But fantasy is not the only way of addressing the problem. That problem also makes a market.

Following Lacan's insistence that "everything we are allowed to approach by way of reality remains rooted in fantasy," future-world building in film may offer visions of masculine fantasies of sexual rapport and satiety, or argue for their impossibility (95). We enjoy the former as a kind of (post-apocalyptic) nostalgia and the latter as (dystopian) frustration, as the infinite foreplay endemic of late capitalism. *Blade Runner 2049* tries to model both.

- - - -

A combination of ecological collapse and a digital information upheaval called the Blackout mark the film's world as different (so far, at least) from our own. Glimpses of big corporate logos for Atari and Pan Am tie the film even more firmly to its predecessor, while subsequently distancing both films from us (as the companies have been consumed or have ceased to operate in our world). Humans and replicants live together, but uncomfortably and unhappily. This isn't yet a world frozen in post-apocalyptic scarcity, but it is certainly cold there. It is close—or far—enough.

In their staging of a future, most post-apocalyptic films and television show us a time in which all the anxiety caused by what author Warren Ellis has called our science fiction condition (*Shivering Sands*) will have been taken care of by rendering life simple and exposed, a time when the tension between desire and satiation may be reconciled, at least for a hero who tries hard. As such, like all fantasy, the average post-apocalypse is conservative and nostalgic in its work to re-inscribe coherent social relationships via recognizable, if antique, racial and economic and (especially) sexual values. The tribalism of *The Walking Dead* and the *Mad Max* films are examples of this, as is the vulgar economic distinctions of *Snowpiercer*, militarization in the television series *Falling Skies*, and the genteel farming and ranching of *The Postman* and the television series *Revolution*. Post-apocalyptic work may be a kind of global turning it off and turning it back on again, a reset in order to duck the problems or brokeness of current social, gendered, economic, and ecological impasses and to re-inscribe simpler—if also often more barbaric—social traditions.

Roughly speaking, in the post-apocalypse, it seems like we enjoy the repeated failure(s) of our own present—climate change, corporate greed, sexual aggression—amplified and projected into remote futures where they will have been addressed, a remoteness that mirrors the remoteness of the past of gladiator movies, fantasy television series, etc. Foundational to these genre worlds is their separation from us now by an obvious discontinuity. In *2049*, the Blackout and the weather are the most apparent breaks, but the corporate logos for defunct businesses work toward this, too, as do the lurid advertising of women. This, it reminds us, is a future that might have been. A future based on a past we have left behind.

It is not too difficult to imagine a serious overlap between the psychoanalytic kind of fantasy and the popular, genre kind of which post-apocalyptic worlds are a subset. Fantasy, both as a genre and as a psychoanalytic category, is conservative and iterative. It is an investment in keeping the world a coherent system of intricately networked actants. Fantasy's job is to turn the repetition of a problem (the impossibility of sexual rapport is a—or *the*—version of such a problem) into a mythic (and therefor distant) resolution. And as fantasy models itself on a received, traditional past, it is often concerned about a fabled and settled beginning that we have lost. As Alenka Zupančič explains:

> repetition exists because there is no linear genesis of the subject (or, to put it the other way around: because there is a subject—the latter being precisely the effect of a dysfunction in the purely linear causality). The genesis of a subject always and necessarily involves a leap, and repetition at its fundamental is repetitious jumping, going back and forth between the edges of this leap. On this particular level repetition is clearly opposed to fantasy, which consists in filling in the gap in question, and transforming the constitutive leap into a linear story. Also, at its most fundamental, fantasy is fantasy about the origins (of the subject). (*Odd One In* 169)

Fantasy is about an origin that would make sense of the mess, tame the possibilities, crystalize relationships, and stop the constant jittering that undercuts any attempt at narrative. For instance, the protagonist of *2049*, K., relies on implanted memories that suggest he has actual parents; "I always told you," Joi says, "you're special. Born, not made." That so much of the film is also concerned with the question of motherhood is perhaps not surprising, either. It is in many ways a film that fetishizes origins, including its own. It is, of course, a sequel.

Additionally, just as individual cosplayers at a convention recreate or riff on an existing commercial property they love (and so eagerly volunteer to advertise for that property), private fantasies are made up of only what is socially available that can be put to work to sustain a consistency of and in the "real" world (in order to allow me to be a functioning, consuming citizen of it). With both kinds of fantasy, following Mark Bould, "the subject is involved in the practice of making do with the texts and commodities that the apparatus provides" (74). This making-do is an appropriation of what is pre-given for personal use. In both, what we enjoy is participation with and in the pleasures that the fantasy offers by assuming its tropes and artifacts as our own. Unless (and even if) you are of a certain age, a 1950s housewife is a memory we have received from elsewhere, from parents or movies or books.

And like the kinds of novels or films that belong to a long or ongoing series, fantasy is invested in elaboration and detail over closure. Obsession with the minutia of staging means not having to see and come to terms with the gap but marks, instead, an investment in telling a believable story and in keeping that story going. In advertising terms, fantasy is about the feature, not the benefit. The obvious, detailed reverence that *Blade Runner 2049* holds for its predecessor indicates the ongoing nature of the narrative. The story matters less than looking and sounding like *Blade Runner*, as evidenced by the interstitial shorts available online that work to fill in the gaps.

It is easy to imagine that Todd McGowan is also describing the completionism of genre-fantasy world building when he claims that, within a psychoanalytic framework,

> fantasy is an imaginary scenario that fills in the gaps within ideology. In other words, it serves as a way for the individual subject to imagine a path out of the dissatisfaction produced by the demands of social existence.... By distorting social reality through an imaginative act, fantasy creates an opening to the impossible object and thereby allows the subject to glimpse an otherwise inaccessible enjoyment. (*Real Gaze* 23)

Just as manic attention to trivial detail endemic to fans of fantasy films and books (and their sequels in any media) demonstrates the amount of our investment in the completeness of a fantasy world, investment in the psychoanalytic brand of fantasy works to "fill the gaps" in the ideological framework of the social "real" world. Both types of fantasy are dedicated to maintaining appearances and to the enjoyment we get from that maintenance. Maintenance here is an investment in narrative and a denial of repetition. It is the spackling of ideology with a wealth of detail that distracts from its shaky foundation. The pleasures involved in main-stream fantasy properties are often found in almost scholarly investments in the details of the worlds being described, in their politics and cuisines and their mating habits that *are* the stories they are telling.

And if mainstream media will not take up this maintenance, fan fiction will and does.

Of course, a world that is very different from our own requires a great deal of detail, of world-building to make it believable. In big stories, difference is often exaggerated to cartoonish effect and goodness is defined by noble, broad gestures. The result is a world in which humans can have little over-all impact (everything has already been killed or is too wild to be tamed by the few humans around). Men are ruggedly heroic, even if they are flawed. Women are in distress, already broken, or simply evil. A single character in one of these stories can often have almost no global influence, but can—through muscles and smarts—radically alter the lives of locals very much like the lone rider from a traditional Western fixes a town. What is re-inscribed is the popular myth of the single great man, a character that is at once aspirational and exceptional.

Through this character, we get to experience a world fixed of its complicated problems by way of a local cruelty and obvious goodness. But, as Kenneth Burke put it, this is "a symbolic getting, not a real one" (272). This character offers us a symbolic way out. He is a fantasy.

In her 2017 McLuhan Lecture, Sarah Sharma argues that the fantasy of exiting is a masculine fantasy that, in certain inflections, is either "a mechanism of establishing sovereignty" or "productive for the politics of maintaining autonomy" ("Exit and the Extensions of Man"). Whether it is for sovereignty or autonomy, I might fantasize about being able to leave the complicated world in which even white men are reduced to commodities, are replaceable, are as disposable as anything or anyone else. The power to exit stands in for every other kind of power, and fantasies of leaving the complications of my world are translated into fantasies of saving a world that better appreciates my "natural" economic, sexual, and racial gifts. I would be a king—maybe replacing an all-enjoying Father—if only.

That the hero of the first *Blade Runner* film—Deckard—is in *2049* found alone among the ruins of Las Vegas makes sense in precisely these terms. His exiting confirms him as a father and his isolation maintains it ("Sometimes, to love someone," he says, "you've got to be a stranger").

Deckard is settled in with empty nightclubs and jittery hologram jukeboxes and fully-stocked bars. Frank Sinatra croons to feed the nostalgia of no one who was alive back then, and really no one at all. Las Vegas has decomposed as the whole world became a market for lowbrow enjoyment. Of course, we are told in both films that those who are best-off have already left Earth. Those who remain make do with a dark world in which entire cities are irradiated, massive e-waste junkyards, or towering apartments. So while the spit and polish of Las Vegas has faded, the rest of the world quit trying to pretend and shambles along, a broken, vulgar world.

It is no accident, I think, that so many post-apocalyptic scenarios rely on the absence or ruins of our usually assumed and invisible infrastructure: the internet, electricity, airplanes, plumbing, even capitalism. We've enjoyed these futures in the original *Planet of the Apes* franchise (especially the first two films) and in movies like *The Omega Man* (1971), *Logan's Run* (1971), *A Boy and His Dog* (1975), *Damnation Alley* (1977), *Escape from New York* (1981), *The Road Warrior* (1981), *The Quiet Earth* (1985), more recent films like *12 Monkeys* (1995), *Children of Men* (2006), *I Am Legend* (2007), *WALL-E* (2008), *The Road* (2009), *The Book of Eli* (2010), *Mad Max: Fury Road* (2015), and television programs like *Jericho* (2006–2008), *Wayward Pines* (2015–2016), and *The Walking Dead* (2010–present). These films and television programs draw attention to what we usually ignore or take for granted, but in a kind of amplified, clogged-toilet way. That is, these generally don't offer critical investigations of the buried, assumed infrastructures that support us daily as much as claim that we're all barbarians once we have to shit outside. Real men, they seem to argue, are what's left once technology is gone.

Of course, the hero of *Blade Runner 2049* flies technology, kills technology, *loves* technology, and *is* technology.

- - - -

Unlike fantasy post-apocalypses, the other, dystopian futures feature the usually invisible social and technical infrastructure that supports our current lives not as absence or ruins or trope-y signifiers, but as the stuff we leave that may live long after us. Just as coaxial cable laid in the 1980s to pipe more television currently connects the internet (and not cable television) to my home computer, what we bury now both makes what comes next possible and ties us up. And, along with laws and bureaucracy (which are also infrastructural), these remains haunt the future. As Steven Shaviro puts it:

> science fiction as a genre does not claim to actually predict the future. Rather, it works to extrapolate elements of the present, to consider what these elements might lead to if allowed to reach their full potential. That is to say, science fiction is not about the actual future but about the futurity that haunts the present. (2)

In this sense of futurity, the things that we have made to satiate us right now survive and inaugurate, if not new desires, at least new ways of figuring our old ones. For instance, with cable television, products available on several channels were just a phone call away. The internet brings home a global marketplace in which we never have to speak to anyone.

But, as Todd McGowan argues, "The only possibility that the future offers is that of a new form of repetition…. When we desire, we look to the future as the time in which we will obtain the object and fulfill our desire" (*Out of Time* 11–12). If *futurity* is a word for the potential of the present, the average dystopia is about how that potential might be actualized. Or, more often, how it won't be.

Recent speculative, vaguely dystopian films like *Her* (2013) and *Ex Machina* (2014) and television programs like *Humans* (2015) and *Westworld* (2016) offer versions of the feminine as products to be enjoyed and as part of infrastructural support for desire. Few feature actual human women, at least in relation to products and their uses. Instead, these films and tv shows often rely on "feminine" tools (usually androids) as surrogates, sometimes as prosthetics, both extending and literalizing the usual post-apocalyptic fantasy move of instrumentalizing women. In this, they beg discussions of fantasy—especially masculine fantasies that make up for the fact of sexual non-rapport—without saying very much at all about women.

There is, of course, a long history of men making women for themselves. Ovid's telling of Pygmalion comes to mind. Or Hoffmann's "The Sandman." Even those scenes in *Vertigo* and *Pretty Woman*, in which

Jimmy Stewart and Richard Gere dress up their real women participate in the theme.

More recently, making women seems to be recognized as an legitimate market—robot strippers are featured at the annual Consumer Electronics Show in our own Las Vegas, Real Dolls are becoming animated, etc.—and market forces do something peculiar to desire. According to Stijn Vanheule:

> Within the capitalistic logic, the lack at the heart of subjectivity is not seen as a structural consequence of using signifiers, but an accidental frustration that can be remedied within the market of supply and demand....
>
> As a result, capitalist discourse implies a particularization of desire, treated as if it is a demand. Whereas in classic [Lacanian] discourse desire is singular in that it cannot be solved by means of the signifier, the capitalist discourse suggests that particular solutions for dealing with subjective division actually exist: the market is there to satisfy customers' demands. (7)

These recent, near-future speculations are very much about market solutions through rendering desire into demand. That these new prosthetics will never be sufficient creates the potential for accessories sales and new models. The repetition that fantasy is supposed to cover or narrativize is translated into supply chains and upgrade cycles that promise movement toward the "better" in their constant iteration.

Capitalism is, at this point, globally infrastructural. It is an infrastructure that we have inherited from the Seventeenth Century. But its latest, global manifestation (what we call *late* capitalism) is a more recent development that led to Lacan proposing a fifth discourse—the Discourse of Capitalism—in the late 1960s and especially in his 1972 lecture at the University of Milan.

This new discourse is a marked departure from the other four. As Colette Soler explains:

> Each discourse...constructs a type of social bond, a sort of standard couple: master and slave, teacher and student, hysteric and master, and

> then psychoanalyst and psychoanalysand.… But there is no such possibility in the scientificised capitalist discourse. It is not a variant of the discourse of the master, and can only constitute a single, barely social bond between the individual and products. Indifferent as it is to "the business of love", it moves towards an increasing fragmentation and instability of social bonds, and leaves individuals always more exposed to insecurity and loneliness. (191)

Each of the four prior discourses—usually illustrated by mathemes—depends on an unconscious or repressed truth that supports those relationships but that is never directly accessed or acknowledged. Instead, this unacknowledged truth is the site that guarantees at least a minimal distance by which a relationship with an other can be sustained. This inaccessible support for the social bond guarantees that whatever message is sent to an other is always never complete or completely understandable. Any social bond is built upon the inability to ever say, or share, it all.

Capitalist discourse subverts that unconscious or repressed truth-support. For instance, in the master's discourse, the unconscious (and therefore unacknowledged) support for the master is the split subject, the fundamental fact of a gap or impasse or impossibility. Ignoring that gap (repression) is what allows the master to take up the position of representative of an established and necessary order. This can be traced back to Aristotle's Mover, but also to the God by whom a monarch receives his divine right, or to the lineage running from the first *Blade Runner*'s Eldon Tyrell to *2049*'s Niander Wallace. The radical shift that Lacan describes as the capitalist discourse is one in which a split, lacking, castrated subject does not address an other *directly*, if imperfectly. Instead, what in the prior discourses would be the unconscious truth supporting the subject's fraught relationship with an other is here opened up in the form of a signifier, a product, an intermediary, through which the subject may access an other that has become, in John Holland's words, a "catalogue of satisfactions" (114).

To return to Soler, the insecurity and loneliness endemic to late capitalism is precisely what is capitalized upon, and is perhaps the most obvious reading of the "social bond" rendered in *2049* as a relationship with Joi, the ubiquitous domestic product with a heart of gold who both has K.'s best interests at heart and is also one of many products in the expansive catalogue of the Wallace Corporation.

The effects of this new availability are double. First, the market insists that a social bond with an other is only now possible via an intermediary object, a product or service that will always focus that rapport in commercial ways. Instead of the misunderstanding that founds a relationship and allows room for fantasy to take hold, this new discourse focusses any exchange through the lens of a product, service, or even capital itself. (Running something—a government or school, for instance—"like it is a business" is precisely the narrowing of the possibilities of social bonds to one relationship via a surrogate that flattens difference and can therefore perpetuate itself across borders.)

And this is the second effect. Once the impasse or impossibility that complicates and slows the social relationship is elided in favor of easy access via a stable, market-friendly key, finding a gap or crack through which to intervene—politically, economically, or psychoanalytically—becomes very, very difficult. The success of the capitalist discourse is that it does not stop writing itself. As Holland describes:

> It reproduces, in the field of the psychic and the social bond, the limitless movement that characterizes capital; both domains are dominated by the same sort of infernal machine. Once the circuit has been traversed and one returns to the beginning at [the split subject], nothing favors one's escape from this discourse and everything leads one, instead, to repeat the same path that has only just been taken. (110)

The trick, though, is that repeating the same path doesn't *feel like* repetition, but iteration. Each new move through the cycle is also to witness a new upgrade, a new feature, a new model that will do it better, without any real critical attention to what *it* might be.

The promise of capitalist discourse is that, in its constant circulation, there is no impasse, no way to acknowledge and traverse the fantasy, no rupture in which we can potentially witness its failure or lack of permanence. Instead, the discourse accounts for discontent within the capitalist system by pushing the final answer perpetually further down the road.

As a subject of capitalism, to the gap about which the subject coalesces is introduced a master signifier (a product) that grants access to the field of signifiers (products, commodities, both here and off-world) that stands for the market as such.

Through this master or introductory signifier (say, Joi), K.'s relationship to the market (the Wallace Corporation, but also the rest of the world that has become a "catalogue of satisfactions") is established and he has access to everything else that the market can offer (upgrades that make Joi portable, for instance). But that market is also evidence that there are more Jois and other models (Mariette, a replicant prostitute, says to K. about his hologram Joi, "Oh, you don't like real girls"). And everything else that is available or just out of stock in the catalogue demonstrates to us customers of the world both the failure of this new particular model to meet every foreseeable need and—regardless of variations, or because of them—the impossibility of any of its contemporaries to completely satisfy, either. The variety upon which a market is founded demonstrates its failure to completely meet the demands of its customers. Yet.

The failure of the market to immediately satiate is, through the constant movement that capitalism enforces, converted into the promise of a future satisfaction only (ultimately) available through participation within the iterative cycle (this feminine companion is good enough right now, but barely), thereby translating desire (which is by definition unsatisfiable) into a demand of and for the future and engendering a subject who can do nothing else but repeat the cycle, since these are now the only terms available to him once he has bought in.

If repetition is the engine of desire, then the engine of capitalism replaces repetition (and desire) with iterative cycles of supply and demand that project the promise of satisfaction into the future, narrativizing the formal impasse as a *not-yet* or *not-quite* or *to-come* such that every failure of every product is also a promise about a future that is already assumed with, in, and by the present. Here, iterative production cycles take the place of singular repetition and demand replaces desire as such.

\- - - -

In many ways—but especially in its treatment of the masculine subject's non-relation to the feminine—*Blade Runner 2049* offers both a near-post-apocalypse and a recognizable dystopia. As such, it both participates in nostalgic versions of masculinity and sexuality that maintain a standard, staid fantasy of completeness *and* it critiques the commodification of those tropes.

As a post-apocalypse, the film insists on a clear, simple, hierarchical order to the world that has been—or will have been—safely re-inscribed in the social order. In contrast to our world, in which even white men are subject to replacement by machines, *2049* extends heterosexist power structures to the order of machines. Following Marshall McLuhan, these machines are in every way an extension of man. Men rule, and if not all men have access to the same power, at least each can count on being the king of his castle through technology like Joi. This is the ideological promise of fantasy. Everyone's place is secured and policed. Everything can be accounted for.

A coherent world is the promise of a lot of popular media, especially movies, and the post-apocalypse may be simply an amplification of what is usual. McGowan describes film's general participation in this promise via a stable, normalizing relationship:

> The romantic union has this ideological weight because it offers a fantasmatic solution to a fundamental social antagonism—that of sexual difference. The heterosexual romantic union that concludes so many films implies that antagonism can be surmounted, that a complementary relationship can be achieved. This idea provides individual subjects with hope that they will find someone to provide what they lack, but it also works to convince them that the social order is a coherent whole (and thus working out successfully). (*Out of Time* 84)

Of course, *2049* does not end with a "heterosexual romantic union." It concludes with the death of the protagonist who, after losing his special love (who is also one among many), sacrifices himself for the sake of the product of the heterosexual romantic union that finished the first film. In this way, as a sequel, *2049* answers the prior film's question *what does it mean to be human?* with iteration itself. That is, it responds with the nostalgic, traditional answer of family and children. But this is undercut by the assumption of sexual reproduction by the corporation as a means of increasing product capacity and meeting market demand, or, in even simpler terms, in the name of expansion and growth.

The parallelism here should not be missed. The nuclear family that is in peril and needs to be saved is exactly the same as the corporate quest for replicant biological reproduction. For both, iteration means the future.

And while the future of *2049* is a pretty recognizable one, that is not because it has much relationship with our own time. At this point, well over thirty years out from the inception of cyberpunk, the genre itself offers little but a nostalgic retro-future that runs almost exclusively on its own tropes and fashions. The film is, itself, an example of the marketing of nostalgia, of fan service that has become the coin of so many attempts at serialized blockbusters and so much television programming. But, at the same time, *2049* does not quite fit, either.

As a product of fantasy and ideology, *Blade Runner 2049* fails to fully satisfy as a sequel or part of a franchise. It is too long, too slow, too loud, and so draws attention to its mechanisms in uncomfortable ways. In terms of "the futurity that haunts the present," *2049* is the dystopian version of mainstream, nostalgic entertainment as it amplifies the common tropes and demonstrates how they work as products, as ways of translating desire into demand for a new episode, and new gizmo, a new partner. These sorts of meditations on genre elements that are usually glossed do critical work in revealing the commonplaces that are marketable and the crowd to which they are marketed. In her defense of the psychoanalytic intervention, Zupančič makes the same point:

> In any social conflict, a "neutral" position is always and necessarily the position of the ruling class: it seems "neutral" because it has achieved the status of the dominant ideology, which always strikes us as self-evident. The criterion of objectivity in such a case is thus not neutrality, but the capacity of theory to occupy a singular, specific point of view within the situation. In this sense, the objectivity is linked here to the very capacity of being "partial" or "partisan." (*What Is Sex* 4)

By over-privileging the usually assumed masculine fantasy of an other to own, *Blade Runner 2049* offers the possibility of critique of masculine fantasy subverted by capitalism *from within*. That is, the prevailing standard of its speculative dystopia is bourgeois, heteronormative, mostly white, and male not (I hope) because we cannot imagine an end to these economies, but in order to show that their failures are *inherent to them* and that the products engendered to cover over these fractures with a fantasy of completeness also highlight a systemic impotence.

References

Blade Runner 2049. Dir. Denis Villeneuve, Perf. Ryan Gosling, Ana de Armas, Harrison Ford. 2017. Warner Brothers, 2018.

Bould, Mark. "The Dreadful Credibility of Absurd Things: A Tendency in Fantasy Theory." *Historical Materialism* 10:4 (2002): 51–88.

Burke, Kenneth. *A Rhetoric of Motives*. Berkeley, CA: University of California Press, 1969. Print.

Ellis, Warren. *Shivering Sands*. United Kingdom: International Electrophonic Unit, 2009.

Holland, John. "The Capitalist Uncanny." *S: Journal of the Circle for Lacanian Ideology Critique* 8 (2015): 96–124.

Lacan, Jacques. *Encore: The Seminar of Jacques Lacan*. Trans. Bruce Fink. Ed. Jacques-Alain Miller. Vol. Book XX. New York: Norton, 1998.

McGowan, Todd. *The Real Gaze: Film Theory After Lacan*. Albany: State University of New York P, 2007. Print.

McGowan, Todd. *Out of Time: The Ethics of Atemporal Cinema*. Minneapolis: University of Minnesota Press, 2011.

Shaviro, Steven. *No Speed Limit: Three Essays on Accelerationism*. Minneapolis: University of Minnesota Press, 2015.

Soler, Colette. *Lacan: The Unconscious Reinvented*. London: Karnac Books, 2014.

Vanheule, Stijn. "Capitalist Discourse, Subjectivity and Lacanian Psychoanalysis." *Frontiers in Psychology* 7 (September 2016).

Žižek, Slavoj. *The Plague of Fantasies*. London: Verso, 1997.

Zupančič, Alenka. *The Odd One In*. Cambridge: MIT Press, 2008.

Zupančič, Alenka. *What Is Sex*. Cambridge: MIT Press, 2017.

7

Object Oriented Subjectivity: Capitalism and Desire in *Blade Runner 2049*

Matthew Flisfeder

Like the original *Blade Runner*, *2049* is a visually stunning depiction of our potential dystopian future; one that if we read it in its historical context provides a detailed cognitive mapping of the continued decline of unfettered multinational capitalism. Also, like the original, the new film provides a surface level portrayal of the world that, if read in spatial terms, maps for us many of the contours of the rhizomatic networks of contemporary capital; however, it also makes plain, thematically, deeper questions about ideology, neoliberal and capitalist subjectivity, dynamics of race and gender, and of course poses questions for us about our new age of automation, global computation, ecological degradation, and provides a glimpse into the vastly developing cleavages between central enclosures and peripheral slums.

M. Flisfeder (✉)
The University of Winnipeg, Winnipeg, MB, Canada
e-mail: m.flisfeder@uwinnipeg.ca

C. Neill (ed.), *Lacanian Perspectives on Blade Runner 2049*,
The Palgrave Lacan Series,
https://doi.org/10.1007/978-3-030-56754-5_7

The film's plot is not overly impressive, nor does it really hit the ball straight out-of-the-park. It is fairly typical in its Oedipal and Christological rendering of its central character, K; and, while the hot takes on the film that have popped up in my social media newsfeed seem to want to stress this avenue of critique, I for one think that there is more to it than simply that, and I have little interest in pursuing this thread. As a product of reboot culture—that ever-increasing number of remakes, sequels, and alternate version storytelling that has become a staple of twenty-first century culture industry—*Blade Runner 2049* was sure to find an audience that would seem to know it all in advance. However, it's my opinion that if we remain focussed on this aspect of the film, then we may end up missing the forest for the anamorphic trees. The film is rich and evocative, not unlike the original; and if I cannot adequately address it all it is only for the better, since *Blade Runner 2049* is much denser than it may first appear. In this regard, here I tease out some of the ways in which *Blade Runner 2049* allows us to reflect upon and register elements of our historical present.

What I find most fascinating about *2049* is its capacity to challenge, not unlike the original film, our notions of human subjecthood. It is by questioning the corporeal and cognitive dimensions of human subjectivity, measuring these against new and contemporary technological developments in Artificial Intelligence (AI), automation, and algorithmic new media, that the film brings to the surface aspects of the present that are cause for reflection and speculation. The film asks us to consider these questions in the context of twenty-first century capitalism; and it does this, as well, by making matters of enjoyment, fantasy, and desire central to our experiences of contemporary selfhood. Or, more specifically, it figures matters of enjoyment, fantasy, and desire as being tied to the "passionate attachments" that constitute us as subjects. It fact, it is when K receives cause to question the presupposition that he is a mere machine that the film provides for us a kind of cognitive mapping of sorts for the cleavages of the contemporary ideology. With K as a model, then, my objective here is to look at the film as a text dealing with questions about subjectivity and desire in the context of the capitalist society. My approach is materialist in the Lacanian-Žižekian sense of positing the subject as the very gap in the Symbolic order, the lack in reality, that opens up the space for an ethical act. *2049* speaks loudly about the differences between bodies and minds of subjects in the world; but as I draw out here, it is by positing the misrecognition of the subject as a gap in reality that we can retroactively conceive the human dimension of *jouissance*, and in this way the film speaks of an "object oriented subject," as opposed to the new materialism of an "objet oriented ontology." But before delving further into the portrayal of the latter in *2049*, please, allow me first...

...A Brief Detour Through *Star Trek*: Does Data Desire?

There is an episode of *Star Trek: The Next Generation* called "The Measure of a Man," in which the agency of Android Lt. Commander Data (Brent Spiner) is put to question. A cyberneticist, Lt. Commander Maddox (Brian Brophy), wishes to take Data apart in order to discover precisely what makes Data "tick"—that is, how he functions so that more Androids can be built to better serve Starfleet. Data refuses the procedure and the issue is then presented: is Data property, or does he have sentience—does he have the "right to choose"? As Captain Picard (Patrick Stewart) is in the process of considering the best defense for Mr. Data, the always wise (and sapient) bartender, Guinan (Whoopi Goldberg), directs him to what's really at stake in this matter: building new Datas is akin to building a new race of beings; and, will not the legal decision about how he is defined—property or free person—come later to affect the treatment of this entirely new race of beings? The question, in other words, pertains to the relationship of master and slave, and the way that this relationship is couched in the legalistic discourses of property vs. personhood; or, perhaps more appropriately, this is a matter that pertains to the cleavage between the object and the subject—between objectivity and subjectivity. Is Data a mere machine, a mere object? Or does he have subjectivity?

The episode mirrors one from the original *Star Trek* series, "Court Martial," in which Captain Kirk (William Shatner) is put on trial for potentially (and intentionally) killing a member of his crew during a dangerous mission. The scenario centres on Kirk's actions and decisions as Captain, and whether or not the accident that supposedly caused the death of a crew member occurred by accident within the context of procedure, or if Kirk failed to follow procedure and is directly at fault. According to his own word, Kirk followed procedure and the death was accidental; however, the computer (machinic) record differs and shows that Kirk *failed* to follow procedure. The episode thus treats the question of human versus machine sentience and the issue of machinic logic, reliability, and objectivity. Liberal society seems to favour what is objective as more truthful or neutral. It's only when we enter the realm of human subjectivity that we discover the existence of a "bias." Like the typical depiction of the surveillance camera, the machine is deemed to show the truth, itself more reliable than the memory of the human subject. But in the episode, the question becomes whether or not a machine can have sentience, consciousness, or if this is a capacity limited to human personhood and subjectivity. Or, since machines do not have sentience, are they not truly objective? Is it human subjectivity that in fact trumps the clout of the machine, or is bias an inherent property of subjectivity? In "Court Martial," sympathies fall towards Kirk and the fallibility of the machine.

Star Trek: The Next Generation further troubles this question, and deals with it in other interesting ways, not least of which is the episode, "Elementary Dear Data," in which Data and Geordi enact a Sherlock Holmes mystery on the Holodeck, where they accidentally create a holographic rendering of Professor Moriarty (Daniel Davis) who develops conscious self-awareness. Although he is but a mere holographic projection, and thus his embodiment is limited to the confines of the Holodeck, the episode takes to task the Cartesian definition of subjecthood, *cogito ergo sum*—I think, therefore I am. Moriarty relates to Data in the sense that he possesses conscious self-awareness and intelligence, but is himself substanceless; Data, conversely has substance but is a product of cybernetics. Thus, in "The Measure of a Man," the question turns towards Data, asking if he, in fact, has a "soul"? But perhaps, the better question to ask is whether or not Data desires.

Star Trek typically draws from and responds to these questions from a particularly liberal humanist perspective of the dualism between mind and body, and attends to them generally in Western legalese. One question that seems to be left unattended when the series questions personhood is the one about desire. Does Data desire? Does he "dream of electric sheep," so to speak? Does he, in other words, have an *un*conscious? In its modern liberal humanist approach to questions of subjectivity and humanity, *Star Trek* takes a more classical stance in placing the conscious self-centred subject as typical. The franchise, however, fails to consider what is perhaps at a deeper level the difference between human and machine: the capacity of human desire and enjoyment, or *jouissance*. My point, in other words, is that the true marker of human subjectivity is less its conscious self-awareness, and more the agency it has in its unawareness or *un*conscious with regards to its own enjoyment. Subjectivity, in other words, emerges, not by directly recognizing and *knowing* what we desire. Subjectivity emerges only after an initial failure—a failure to obtain the object that we *believe* we desire. It is our *un*awareness—our misrecognition—of the fact that we are following a different path than the one that we assume that accounts for the agency of the subject. Recognition is merely a *lure* that leads us in the direction of our actual (unconscious) enjoyment. Here, *Blade Runner* is more precise.

Substanceless Subjectivity: Are We Human? Or Are We Denser?

Much has already been said or written about the question of Deckard's (Harrison Ford) personhood.[1] When, at the end of the original film—or more adequately in the 1992 Director's Cut and the 2007 Final Cut of the film—Deckard finds the origami unicorn left by Gaff (Edward James Olmos), he gets an indication that he too is a Replicant. The paper unicorn mirrors the image of the unicorn that we see in Deckard's dream sequence inserted into the film in the Director's Cut at about the midway point. Throughout the film, both Deckard and the audience are meant to be misled about his status as human. He is apparently a human blade runner, whose job it is to hunt down and "retire," or kill rogue Replicants deemed illegal on Earth following a mutiny of the Replicants on one of the Off-World Colonies, and the mass murder of the human "masters" living there. As the original film explains, affect became a factor in losing the ability to control the Replicants. The NEXUS 6 model Replicants started to feel their own emotional responses, and in order to help tame these responses the manufacturing Tyrell Corporation opted to give the Replicants implanted memories that could help them to better assimilate affect into their cognitive relationship to reality. Affect and memory make the Replicants, like Rachael (Sean Young) and Deckard (hypothetically NEXUS 7 models), "more human than human," as the motto goes for the Tyrell Corp. Ultimately, it is the inscription of fantasy, and not simply memory, that tricks the NEXUS Replicants. It is their relationship to a fundamental fantasy that humanizes them, by providing for them a relationship to their enjoyment. But this is where the new sequel, *Blade Runner 2049* complicates matters even further.

The new film takes place thirty years after the story in the original. The opening to *Blade Runner 2049*, and subsequent expository details brought out through dialogue between the characters, explains that things have changed since the time of the original film. This background is further developed in a series of three short films produced as side promotional projects for the new film: *Black Out 2022* (Dir. Shinchiro Watanabe 2017), *2036: Nexus Dawn* (Dir. Luke Scott 2017), and *2048: Nowhere to Run* (Dir. Luke Scott 2017). *Black Out 2022* explains that as the NEXUS 6 inventory expired, the Tyrell Corp. produced NEXUS 8 models that were purpose-built and with natural lifespans. The NEXUS 8 models were integrated into human society on Earth. However, disdain towards the Replicants persisted resulting in the rise of "Human Supremacy Movements," which used Tyrell's Replicant records to track down and kill them. In retaliation, a Replicant liberation movement destroyed all of the datacenters housing the records, causing a black out that swept across the world.

The blackout is referenced several times throughout *Blade Runner 2049*. It led, afterwards, to the prohibition of Replicant production, which forced Tyrell out of business. However, the second short promotional feature, *2036: Nexus Dawn*, depicts how fourteen years after the black out, and the prohibition of Replicant production, the new tech mogul, Niander Wallace (Jared Leto)—who the opening scroll sequence at the beginning of *Blade Runner 2049* explains built his fortune by inventing a new protein harvesting technology to stave off the erosion of the food supply and the hunger that followed the blackout—has bought out the Tyrell Corp. and has begun to again produce new Replicants, which he has now made to directly obey the human masters. It's this fact that allows Wallace to force the future government to rescind the prohibition and the new mass production of Replicants.

In *Blade Runner 2049*, unlike the original, Replicants are now allowed to roam the streets freely. They are permitted because of the failsafe of obedience. The film's hero, K, is a new generation NEXUS. Like Rachel (and Deckard), K's memories have been implanted. But unlike her, K is made fully aware that he is a Replicant and that his memories are fake. In the original film, Deckard tells Rachel that she is a Replicant. Rachel responds by showing him photographs—"objective evidence"—of her childhood, of her childhood memories. These photographs represent elements of her personhood measured in memories; but they are memories in object form. Deckard tells her that these memories are in fact fake, that they are images of Tyrell's niece. Rachel's memories, that element of herself that made her "human," actually belong to someone else. It's this fact that subtracts for her the essence of her personhood; afterwards she devolves into a kind of substanceless subjectivity. It's this fact that, in a sense, drags her down from the highest to the lowest; it is her process of proletarianization, so to speak. Her photographs, on the one hand serve as objectal correlates to the fundamental fantasy that supports her existence as human. But, on the other hand, the film shows that even the objective evidence of her personhood is without guarantee. It is not the object, but her experience as substanceless that transforms her into a subject. She is "hystericized" (in the Lacanian sense) at the moment of the failure of her fundamental fantasy, forcing her into a position of questioning her identity in the eyes of the big Other.

Although K is aware that his memories are implanted and that he is a Replicant, his process of proletarianization occurs somewhat differently. Like the original film, *Blade Runner 2049* maintains a focus on questioning liberal conceptions of humanism. And similar, as well, to the original film, the new one hinges its conception of humanity on the difference between conscious self-awareness and misrecognition, unawareness (of what we really are), and an "unconscious" of sorts.

Whereas Rachel believed herself to be human, only to discover that she is actually a Replicant, K's mystery begins when it appears as though he might not in fact be a Replicant but human—or, at least, human-like. K and Rachel are different from a character like *Star Trek's* Commander Data for the fact that their subjectivity is defined, not by conscious self-awareness, but by conscious misrecognition—the misidentifying, mis-interpellation, of the core of their subjecthood. This possibility presents itself to K in the process of investigating a Rogue NEXUS 8, Sapper Morton (Dave Bautista).

Blade Runner 2049 begins as K tracks down and "retires" Sapper—in scenes that mimic the aesthetic and the staging of Deckard's fight with Roy Batty (Rutger Hauer) in the Bradbury building at the end of the original film. Afterwards, while investigating deeper into Sapper's protein farm, K discovers a box buried beneath a tree, which has carved into it the numbers "6 10 21." Upon seeing these numbers, a memory sequence is flashed of a small child holding a small wooden toy horse, standing in front of a fire—K's fundamental fantasy. The sequence is flashed quickly and becomes the film's McGuffin (of course, the toy horse replicates the origami unicorn left by Gaff in the original film). It's later revealed that the contents of the box are Rachel's remains. It's discovered that she died not long after giving birth to a child delivered through c-section. This astounding revelation is the political linchpin of the film.

Thought to be impossible, the fact that Rachel was able to give birth to a child raises concern for K's superior, Lieutenant Joshi (Robin Wright). She is concerned that if the Replicant resistance movement discovers that a Replicant gave birth to a child then there would be a new war, a new battle cry for the Replicant resistance. She orders K to find and kill ("retire") the child. But K still has an odd connection to the site from which Rachel's remains were discovered: the memory that was triggered by the numbers carved on the tree trunk. What is his connection?

The numbers carved out on the tree give material substance—not unlike the origami unicorn—to his memory. It objectifies, in a way, a memory that he had until then believed was implanted. Although K knows that he is a Replicant, and knows that his memories are implants, the material manifestation of this memory troubles his established subjecthood—perhaps he is not just a mere Replicant. Perhaps he is actually Rachel's child. Is *he* the child that Joshi has ordered him to kill? If a child is born from a Replicant mother (or parents), does he remain a Replicant? If he has produced his own memories, is he still a Replicant? What is now the dividing line between humans and Replicants if the latter can self-reproduce? What marks our humanity?

The fact that K is also aware of his existence as a Replicant (unlike Rachel and Deckard), and that he knows that his memories have been implanted, presents an intriguing update to *Blade Runner's* universe of ideology critique. In contrast to an older notion of ideology as false consciousness, K operates less through the mechanism of mystification, and more through that of disavowal as in the psychoanalytic statement, "I know very well, but nevertheless..." His obedience is predicated upon a certain disavowed knowledge, where in the place of this knowledge he is allowed to enjoy. Are things really so different for the way that we humans in the real world contend with our everyday existence, given the contradictions of neoliberal capitalism, of which we are constantly made fully aware? Or is subjecthood in contemporary capitalism not also predicated on the same kind of disavowed knowledge. What then happens when the *misrecognition* central to our conscious subjecthood is brought into question?

Automation, Replication, and Enslavement

The political economic context of *2049* is also brought into focus through the contradictory perspectives of two secondary characters. Whereas Lt. Joshi fears the possible repercussions of the discovery that Rachel, a Replicant, gave birth to a child—that if discovered, this might cause a new war—Wallace is oddly excited. From his perspective, as a capitalist and owner of a megacorporation, the self-reproduction of Replicants might actually help to solve a problem of production—that is, the speed with which he is capable of turning out new models to satisfy demand from the Off-World colonies. While the state political-legal position of the LAPD is one of potential conflict, Wallace sees only the revolutionary productive potentials of self-reproducing Replicants, which he hopes could give him a leg up in his business. His perspective is one of the market; and it is worth looking at these contradictory perspectives of Joshi and Wallace, for it is indicative of the contradictions that do exist between the political and the economic; or, put differently, it oddly indicates the intersection of the class state mechanism and the tensions in the economic mode of production.[2]

Wallace, a capitalist, is continuously interested in driving further the revolutionizing of the means of production. If we read this from a traditional historical materialist perspective, we might be able to register this drive as one that amounts to escaping or overcoming the contradictions inherent to the mode of the production. What are Replicants if not the means of production outgrowing the relations of production? They are both the conscious operators of production *and* the means of production—all of the forces of production congealed into themselves. They are akin to the drive to automate the production process so that capital can retain more profit that does not have to be distributed in the form of wages to human labour. Wallace is (literally, in the film) blinded to this as he strives to further advance processes of automation located in the Replicants themselves, as productive technology. They are from the perspective of the Corporation (both Tyrell and Wallace) non-human labouring machines, or forms of labour-saving technology; machines that take on the appearance—they look like—human slaves. So what happens when the slave-machines begin to recognize themselves as sentient beings of rights? This is the question of race and slavery that Guinan posed in *Star Trek*.

One of the contemporary trends in philosophy is the so-called New Materialism, or Object Oriented Ontology. In a recent article that is well worth some reflection, W. Oliver Baker makes the claim that interest in the New Materialisms is very much a particular reaction to the current moment of capitalist development (Baker, 2016). Within the history of colonialism, Baker notes that colonized peoples were not in fact viewed as human, but as non-human objects. Slaves, for instance, are considered to be productive *property*, means of production that are more like machinery than people. Contemporary neoliberal capitalism has radicalized this situation somewhat; that is, if we go to the end in considering the effects of the neoliberal rhetoric of entrepreneurialism.

For the neoliberal ideology, democratic citizenship has been replaced by entrepreneurship. We are now all "entrepreneurs-of-the-self." This has emerged in material practice as the austerity logic of neoliberalism has torn down the social safety net, making the market the ideal space from which access to needs is provided at a cost. Sure, we are all free to access our needs in the market, but rising costs of real needs effect an imbalance that reproduces inequality. At the same time, just-in-time models of production, and an increase in precarious labour, enforce new *regimes* of labour where every entrepreneur-of-the-self is constantly in the process of both producing and reproducing selfhood (what Foucault calls "biopower") at the same time that they are also producing value-bearing commodities. Michael Hardt and Antonio Negri read this process as one of "subjects producing subjects;" (Hardt and Negri, 2000: 32) but as I have argued previously, this is a process not of subjects producing subjects, but of subjects actively involved in the deepening processes of reification that now occupy the entirety of life (Flisfeder, 2015). This is true of Foucault, as well, who, as I've put it, "represents the individual caught in ideology insofar as it misrecognizes its objectivization as a condition of its subjectivization" (ibid.: 568). In other words, instead of producing new subjectivities, my claim is that neoliberalism is a term that helps to justify contemporary processes of deepening reification; and, back to the New Materialisms that Baker addresses, it seems as though something like Object Oriented Ontology begins to factor in at the moment when the colonialist human begins descend into the sphere of the objectified non-human commodity. More than just relations between objects replacing relations between subjects—or, in terms of commodity fetishism, where a political relationship between people is masked by a social relationship between things—now we begin to conceive a horizontal network of objects that includes human actors that are somewhat akin to non-human agents. But only at this moment of total reification, does something like Object Oriented Ontology, as Baker argues, seem to gain traction. So, back to *Blade Runner 2049*: what happens when the objects, the machines, begin to gain sentience, when the machine-objects are subjectivized?

This process of subjectivization is more complicated than in the older Althusserian sense of ideological interpellation, which the Slovenian school of Lacanian scholarship has troubled for years. Whereas for Althusser, the subject is a product of ideology, according to Mladen-Dolar, the subject is what emerges where ideology fails (Dolar, 1993). This makes subject, then, not a category of ideology, but one of emancipatory agency, and demonstrates effectively where a reified consciousness differs from that of the proletariat. This, we might say, is what happens to K upon questioning the possibility of his status as Replicant. It's when he begins to doubt his existence as a normal Replicant that he becomes an hystericized subject in the Lacanian sense of demanding from the big Other: *Que vuoi?*—What do you want from me?

Jo(u)i-*ssance*

The object quality of Replicant subjecthood is further taken to task in *2049* through the character of K's holographic girlfriend, Joi (Ana de Armas). Joi, too, is a product of the Wallace Corporation, but unlike the *embodied* Replicants, she is a hologram. Of course, her depiction in the film abounds in not so subtle references to recent and past depictions of "robot" girlfriends, from 80s films like Steve De Jarnatt's *Cherry 2000* (1987), or even Michael Gottlieb's *Mannequin* (1987), to more recent versions in films like Alex Garland's *Ex-Machina* (2014) and Spike Jonze's *Her* (2013), and TV shows like *Westworld* (HBO 2016–). Joi is bound to her holographic projection unit built into K's apartment and therefore unable to be truly mobile. She only becomes mobile when K buys her a "gift," a portable holographic projection device that allows her to be transported to other environments, outside of K's apartment. Although it seems to give her freedom, we see throughout the film that she can only go as far as K, which is to say that her "freedom" is still tethered to K's mobility and not her own.

Joi serves somewhat as K's counterpart, an artificial lifeform whose sentience is continually put into question. More so, in some ways, than even K since as his stereotypically domesticated "housewife," her portrayal raises questions about the gender dynamics and politics internal to the identity of the Replicants. Because she is without material substance, it is depicted in the film that Joi and K are unable to engage in physical acts of love (kissing, holding hands, and having sex). This impasse allows us to reflect upon the depiction of some of the other female characters in the film.

Mariette (Mackenzie Davis) is a prostitute who, at an earlier point in the film, is told by Freysa (Hiam Abbass), a leader in the Replicant resistance, to attempt to seduce K—likely to bring him to the resistance. Later on, Mariette arrives at K's apartment at Joi's request. Since Joi and K are unable to have physical sex, Mariette is brought to serve as Joi's surrogate, replicating a popular scene from Spike Jonze's *Her* (2013), in which the Operating System (OS), Samantha, with whom the film's hero, Theo, has fallen in love, invites a human "surrogate," Isabella, to act as a material stand-in for them to have sex.[3] The scene in *2049* has Joi's holographic image overlaid and synced with Mariette's body, which can be used to satisfy K's pleasure, while continuing to fantasize about having sex with Joi directly. Both Joi and Mariette (whom we also later find out is a Replicant, part of the resistance), then, here appear to serve as objects—in both physical and virtual form—for K's enjoyment. They serve as the props of his masculine sexuation, his identification as a masculine subject, in the way described by Lacan in his logics of sexuation (Lacan, 1999). We might even say that Joi serves as the fantasmatic, sexualized counterpart to the memory of the horse toy that plagues K from the beginning.

The scene is unambiguous in its depiction of the heterosexual male fantasy, but I would also argue that because this is so overt it is meant to trouble even the continued existence of sexism in contemporary capitalism. The film appears to be quite self-conscious of this fact as the villainous Replicant, Luv (Sylvia Hoeks)—Wallace's assistant and henchwoman—twice asks K how he likes their "product." This ironic and doubly-coded depiction of sexist gender dynamics is also evinced in the way that Joi insists on giving K the name "Joe." K confides in Joi that he believes himself to be Rachel's child, to which she responds by telling him that she's always felt that he was special, insisting that instead of his Replicant name, K (a shortened version of his serial number), he uses the human name to identify his unique qualities amongst the Replicants. Only, later on in the film, K wanders through the L.A. streets and comes upon an interactive pornographic advert for the Joi product, which towers over him, leaning in to him saying: "You look like a 'good Joe.'" It is clear in this instance that the amorous feelings that Joi seemed to be projecting onto K were still part of her programming. It also becomes clear at this point, through this realization, that K has in some ways traversed the fundmental fantasy that was his tether to his own social being.

The female Replicants still carry with them the object-qualities addressed above in relation to the question of non-human slaves; and even Joi's immobility is an indication of this, comparable to the position of the memory designer, Dr. Ana Stelline (Carla Juri). K seeks out Stelline to discover whether or not his memory of the child is real or an implant. She is, herself, confined to a cage-like environment, quarantined due to a rare illness that she has acquired. She tells K that it is illegal to give Replicants human memories, which we can deduce has changed since Rachel as a model 7 NEXUS. Stelline reads K's memory and tells him that the memory was in fact lived by a person, which makes K believe finally that he *is* Rachel's child. When we later discover that Stelline and not K is Rachel's child, this further complicates the ethical position of the film. The memory of the horse toy belongs to Stelline. It was implanted into K's memory. Although this reconfirms his status as a Replicant, the process of searching out the object still helps to place him in the position of the ethical subject, not unlike the process of the analytic treatment. After going through the process, after searching out the object and ending up still with nothing—he is not Rachel's child, he is still formally, technically, the same as he was before—the process still allows K to return to his previous objective position, but now with a newly formed subjective perspective on his ethical position. He has lost Joi (joy/pleasure), but he has traversed the fantasy that positioned him within the existing relations of power. He, in other words, did not give way to his desire, and ended up following this path long enough to reconceive his relationship to his en-joi-meant.

Putting this last point aside for now, it is worth comparing Dr. Stelline's and Joi's entrapment and confinement to their respective camps since what the film also provides for us is cogent investigation of the politics of space and how this represents key elements of the world that the film draws out. It creates for us, in a sense, a spatial representation of the rhizomatic architectures of internalization and externalization that are part of global late capitalism, but which are also indicative of the wide array of global spatial conundrums and contexts that we now face, from ecological catastrophe to the networked aspects of global production and to the totalization of global computation that is now a part of our everyday experiences of reality.

The Camp or the Bunker

Another way that *2049* differs from *Blade Runner* is in its depiction of the world outside of the dystopian Los Angeles. The original 1982 theatrical release of *Blade Runner* gives us a small utopian glimpse at the world beyond the dark L.A. borders, when Deckard and Rachel escape up north. This ending is of course removed in the Director's and subsequent cuts of the film, leaving us only with the claustrophobic space of dystopian L.A. The only indication of a place beyond comes from the overhead floating adverts for the Off-World Colonies. But in *2049*, we do get to see a broader scope of the planet.

The film opens, even, with the scene at Sapper's protein farm, which appears dusty, barren, and cold. K later goes off to find Deckard who is hiding in a destroyed Las Vegas, which has been polluted by radiation, making it impossible for any humans to live there (since both Deckard and K are able to withstand the radiation, we can assume that this is a subtle indication that they truly are both Replicants). Such depictions thus force us to consider L.A. as a space of both enclosure and respite. In *Blade Runner*, the constant adverts for the Off-World Colonies make it seem as though the inability to escape Earth is a fate left to the misfortunate lower classes. In *2049*, the extreme ecological degradation in the various other parts of Earth present Los Angeles less as a camp than as a bunker, a space of protection from the dangerous world beyond its borders. But the city as centre also depicts for us in another way that key element of disavowal that is a component part of contemporary ideology.

We see in the process of K's investigation, when he attempts to track down the origins of his memory of the child with the small wooden toy horse standing in front of the fire, the existence of non-machine labour. That is, we are given here a scene depicting the production of the machinic through child slave labour, which we can clearly recognize as akin to modern day sweatshops, often set on the periphery of the capitalist global north—which is to say in the impoverished developing locales of the global south, officially condemned by liberal capitalism, but still an elementary part of the capitalist pursuit of increasingly cheaper labour. For all of the glitzy glossy neon and spectacle of the big city, it's very construction is still reliant upon various forms of racialized, gendered, and child exploitation. Fortress L.A. is the very shield, the bunker that protects us from the hard reality of the world beyond.

But we know this already. Who doesn't know about the existence of global ecological degradation? Who of us remains unaware of continued existence of racialized and gendered practices of exploitation—in both the developed and developing worlds; or of the continued existence of child exploitation in sweatshops. We know, but nevertheless… In this way, we are all a bit like K, consciously misrecognizing our own subjective position in the world; and, in this way, too, *Blade Runner 2049* reflects perfectly the now.

Just as K's initial misrecognition leads him towards an emancipatory logic, so too does the narrative of the film express this possibility for us under the present conditions of contemporary capitalist ideology. Like K, we—in twenty-first century, postmodern and neoliberal capitalism—relate to the present ideology through a form of misrecognition and disavowal. K begins by *knowing*—that is, by "recognizing"—the fact of his existence as a Replicant. He is able to consciously acknowledge that he is a Replicant with implanted memories. But he disavows this fact as part of his daily existence. Today, we all know the fact of capitalist exploitation, we are able to avow this as a recognized fact; nevertheless we act as if this was not the case. K begins his emancipatory journey when, for an instant, driven by the logic of his desire, he begins to *doubt* his existence as a Replicant. Driven by this possibility he seeks out the truth. He misrecognizes the logic of his desire, but it is only through such a misrecognition that he ultimately becomes subject. In the end, he returns to his initial position of existence as a Replicant, but now with a new understanding as to his *conditions* of existence and possibility.

As the logic of the film suggests, we become politicized, "not out of some neutral concern for larger political questions or some universal desire to eliminate injustice but because of a singular desire that bear's only on one's own subjectivity" (McGowan, 2011: 115). This is the lesson of the film: human subjectivity bears upon our relationship to our desire, and the political ethics of the subject are the same as the ethics of psychoanalysis: do not give way with regards to your desire, even if, in the end, we gain nothing but a new perspective on the very position from which we began. In *Black Out 2022*, the timid Replicant character, Trixie, asks her savior, Iggy: "if we die, we go to heaven?" He responds: "no heaven or hell for us; this world is all we've got;"[4] and, perhaps, in the face of all of the various conflicts and struggles that we anticipate facing in the progression of twenty-first century capitalism, of which we are already fully aware, this is, ultimately, the materialist point of *Blade Runner 2049*.

Notes

1. For more on this, see Matthew Flisfeder, *Postmodern Theory and Blade Runner* (New York: Bloomsbury, 2017), 121–135.
2. This, still, was present in the original film; as Ian Buchanan indicates, what *Blade Runner* stages "is a confrontation in which the ingenuity of capital... is pitted against the vigilance of the State." *Deleuze: A Metacommentary* (Durham, NC: Duke University Press), 132.
3. On this, see Matthew Flisfeder and Clint Burnham, "Love and Sex in the Age of Capitalist Realism: On Spike Jonze's *Her*." *Cinema Journal* 57.1 (2017): 25–45.
4. Which I'd like to contrast with Father John Misty's point at the end of his song, "Pure Comedy:" "I hate to say it, but each other's all we've got."

References

Baker, O. W. (2016) "'Words Are Things': The Settler Colonial Politics of Posthumanist Materialism in Cormac McCarthy's *Blood Meridian*." *Mediations* 30.1: 1–24.

Dolar, M. (1993) "Beyond Interpellation." *Qui parle* 6.2: 75–96

Flisfeder, M. (2015) "The Entrepreneurial Subject and the Objectivization of Self in Social Media." *South Atlantic Quarterly* 114.3: 553–570.

Hardt, M. and Negri, A. (2000) *Empire*. Cambridge, MA: Harvard University Press.

Lacan, J. (1999) *Encore: The Seminar of Jacques Lacan Book XX. On Feminine Sexuality, The Limits of Love and Knowledge, 1972–1973*. Ed. Jacques-Alain Miller. Trans. Bruce Fink. New York and London: W. W. Norton.

McGowan, T. (2011) *Out of Time: Desire in Atemporal Cinema*. Minneapolis, MN: University of Minnesota Press.

8

What Happens When the Replicants Become Extimate? On the Uncanny Cut of the Capitalocene in *Blade Runner 2049*

Alexander Bove

> Anxiety is this cut—this clean cut without which the presence of the signifier, its functioning, its furrow in the real, is unthinkable—it's the cut that opens up, affording a view of what now you can hear better, the unexpected, the visit....
>
> —Lacan, 76

> The uncanny is always at stake in ideology—ideology perhaps basically consists of a social attempt to integrate the uncanny, to make it bearable, to assign it a place.
>
> —Mladen Dolar, 19

Blade Runner 2049 revolves around a kind of anxiety point, a cut or breaking point between what we know and something unexpected. It envisions a point where something emerges into the world which is a cause of anxiety over the structure of the world as such, a signifier that disrupts the world's ontological consistency ("this breaks the world" as

A. Bove (✉)
Pacific University, Forest Grove, OR, USA
e-mail: above@pacificu.edu

C. Neill (ed.), *Lacanian Perspectives on Blade Runner 2049*,
The Palgrave Lacan Series,
https://doi.org/10.1007/978-3-030-56754-5_8

Joshi puts it), and the story that unfolds is structured by an attempt to manage the relation to it, to capitalize it, control it, frame it. What is *Blade Runner 2049* about but the struggle for power that comes from assigning a place to the uncanny—the Thing that cuts across our symbolic categories of self and Other—and the attempt to resist this power struggle?

In a scene set in the futuristic LAPD headquarters, the anxiety-point of the cut emerges as something hidden that comes to light: a threat to the humans' ontological order that sparks the race for power and control central to the film's plot. Fittingly it is only visible through computer-mediated digital enhancement to a nonhuman subject. Officer K, a replicant Blade Runner, has found something unexpected while on a routine mission to "retire" a rogue replicant: beneath a propped up dead tree, buried under the infertile ground of the now-inhospitable earth, K located a concealed box that turns out to contain some bones. Back in the police department lab, we find that these are a woman's bones, a woman who has died in child birth, which is testified to by a cut in the pelvic area, apparently from a C-section rudely executed with a razor by the replicant K has just retired. The human operating the computer-enhanced images can see no further, but K, being nonhuman, sees something beyond the scope of the human eye, there at the level of the cut that marks an "impossible" birth there is a serial number, a signifier that reveals something unthinkable: the nonhumans created by humans have begotten a (non)human.

Why, then, is the anxiety point of *Blade Runner 2049*'s world structured around *birth*? Or as Žižek puts it: "The first question to ask is: Why is the fact that two replicants (Deckard and Rachael) formed a sexual couple and created a human being in a human way, experienced as such a traumatic event, celebrated by some as a miracle and castigated by others as a threat?" In what follows I would like to explore the ways in which *Blade Runner 2049* structures and navigates our relation to a particular anxiety, or trauma: the anxiety over the essence of the human, yes, but more specifically, the anxiety surrounding the internal/external division between the human and its own nonhuman surplus, or the cut of the *real* in Lacan's terms, as well as the *jouissance* associated with this anxiety point. The name psychoanalysis gives to this particular form of trauma/anxiety/*jouissance* is the uncanny. We could say, then, that *Blade Runner 2049* differs from the original *Blade Runner* in it's fundamental concern, which it shifts from the existential question of *what it means to be human* and how nonhuman objects like machines can simulate or recreate that "essential" x-factor, to the now more timely question of the (uncanny) status of the ontological cut from the Other (for example nature) that sustains the idea of the human and human culture as such, a question belonging to a new phase of late capitalism often referred to as the anthropocene or the capitalocene. In other words, *Blade Runner 2049* is more of a Frankenstein story, a story of the creation of a creature that can procreate—that is, provided we remember that Mary Shelley's *Frankenstein* is essentially not about the manufacture of a particular living creature, but rather the human desire for dominion over life itself, nature's "ultimate mystery," in the creation of a new *species* from inert matter, a therefore "perfect" species which would fill an ontological gap between humans and nature, subjects and objects. As Mladen Dolar points out, *Frankenstein* is therefore about a "quest for a 'zero degree' of subjectivity" (17).

The Cut of the Real and Zero Degree of Subjectivity

Freud's concept of the uncanny comes at a crucial juncture in Freud's *oeuvre*: in brief, it marks a point at which his model of the human psyche is disrupted from within by an external foreign entity: the death drive, which is hinted at in his essay "The Uncanny" (1919) through the concept of *repetition compulsion* and developed explicitly in his subsequent work *Beyond the Pleasure Principle* (1920). In his seminal essay, Freud defines the uncanny as a play between the familiar and the strange—or as Freud puts it, "the secretly familiar"—the homely (*heimlich*) and the *un*homely (*unheimlich*), which arises with the return of the repressed; thus the uncanny describes the feeling when a forbidden/repressed/ secretly known thing "comes to light." As a liminal concept wavering or flickering between oppositions—strange/familiar, self/other, animate/inanimate—yet reducible to neither side, the uncanny is a notoriously elusive category. Yet this strange *undecidability* between inside and outside touches upon one of the basic ambiguities of the term that will help in understanding the uncanny question underlying *Blade Runner 2049*: why the ontological emphasis on a subject being created *within* a body as opposed to *outside* it, or, in other words, on what humans call *birth*?

In fact, in Freud's essay, the return of the repressed is always essentially traceable to a moment in the history of the subject that logically precedes the relation between inside and outside or self and Other. Both the "castration complex" and "primary narcissism," the two forms of repressed material most associated with the uncanny, refer implicitly to a *cut* that establishes a relation between inside and outside of the subject's body and psyche, but which, precisely because it plays the role of positing this structure in the "first place," as it were, always threatens to "return" the status of the (now repressed) non-relation. As Freud puts it, the different examples of the return of the repressed he provides as examples of the uncanny all "[harken] back to particular phases in the evolution of the self-regarding feeling, a regression to a time when the ego had not yet marked itself off sharply from the external world and from other people" (Freud 212). What is touched by the uncanny, we could say, always bears the mark of this cut, this (impossible) known within the unknown, the foreign and strange within the familiar, that threatens to undue the very distinction between the separate categories, which is why Freud's tracing of the uncanny to castration and narcissism is always already a mark of *symbolic* castration; looking forward, I will call it an uncanny *desubjecitification* that always haunts subjectification.

But what if subjectification does not apply to a *subject*? Can a nonhuman *object* experience the uncanny? When K finds a number inscribed on a tree on Sapper Morton's farm he experiences it as *secretly familiar* because it immediately raises questions concerning the distinction between interiority and exteriority: *why is my memory, which I also "know" to be a manufactured fiction implanted within me, reappearing to me externally in reality, outside in the world, in this unfamiliar place, however unexpectedly homely it may appear*?

The dimension of the uncanny here should be sharply distinguished from the simulacra effect often translated in the terms of the now trendy "uncanny valley" motif: the potential uncertainty in telling between a human and an automaton. Matthew Flisfeder, for instance, made an important observation when he points out that in *Blade Runner 2049* this question of the relation between humans and replicants is actually brought to the level of a more fundamental, and perhaps contemporary, question: what is the relation between subjects (humans) and objects (nonhumans) not only in the form of computers or AI but as machines or equipment, as evidenced in the equating of slaves and agricultural equipment by American slaveholders and in the exploitation of laborers under capitalism.[1] For Flisfeder this reconfiguration of the ontological question/dilemma reflects the social conditions of modern capitalism that objectifies human beings based on race, class, gender, etc. Although Flisfeder doesn't draw any connections to the uncanny here, I would add that looking at this question through the frame of the uncanny highlights the complexity and immanence of its historical dimension, it's immediate reference to our world today, to what has become know not just as late capitalism but as the *capitalocene*.[2]

In the capitalocene, then, our relation to the other, to nature, and to the nonhuman, is inextricable from our relation to objects. In this sense, rather than ask, *what is the difference between a human and a lifelike android?* we should perhaps ask, *what is the difference, in the ontological sense, between a human and a glass of water*?—or any object designated as such in ontological opposition to my subjecthood. Thus Flisfeder notes the popular turn towards objects in recent forms of "New Materialism," especially in the form of Object Oriented Ontology (or OOO), which he sees as illustrating this trend's relation to neoliberal capitalism: "instead of producing new subjectivities, my claim is that neoliberalism is a term that helps to write off the processes of deepening reification; and... it seems as though something like [OOO] begins to factor in at the moment when the colonialist human begins to descend into the sphere of the objectified non-human commodity." One can see how the concept of reification comes quickly to minds as Timothy Morton, for instance, one the major proponents of OOO, considers conscious subjectivity as an effect, a kid of "realist illusion," of ontology, making the subject nothing more than a particular *type* of object. In this way, we could say that Morton's version of OOO is uncanny, or sees objects as uncanny

("Objects are not just themselves—they are uncanny: they are both themselves and not-themselves," Morton 64) in a way perhaps comparable to Psychoanalysis, the uncanny science that historically overturned or decentered our concept of what a subject is.

So how can OOO and psychoanalysis be compared, and how is the concept of the uncanny different in each? What ultimately allows for the uncanny play of objects that subverts traditional metaphysical distinctions like active/passive, subject/object, argues Morton, is the ontological "Rift," intrinsic to all objects (whether animate or not), "between *essence* and *appearance*."[3] This rift, the fact that essence always recedes beyond appearance, gives objects a "split" structure of "self-withdrawal," as Levi Bryant calls it in *Democracy of Objects*, that prompts Bryant to compare it to the structure of the Freudian/Lacanian split subject. Despite Žižek's objection here that "if there is a self-withdrawal, there has to be a Self from which its own substance is withdrawn—and one cannot in any meaningful sense call the actual relations of an object to other objects this object's self" (66), Timothy Morton *does* see objects as having egos, memories, and even dreams. Taking his cue from Freud's discussion of the ego as an object cathexes, that is, from the idea that the ego can be treated like an object by the unconscious or even by itself, Morton explains:

> If ego is object-like, then the inverse surely applies. The identity of this glass is the way I use it as a glass by pouring water into it, and the way it was shaped as a glass. And again, there is a profound Rift between the identity of the glass and the essence of the glass….[I]t's *the difference between the glass and the glass*…. The glass is a glass and an uncanny not-glass. (212)

But being *treated* like an object, or being "object-like," is not ontologically the same as *being* an object; and here we can point to the fundamental difference with respect to psychoanalysis, especially Lacanian psychoanalysis. Morton, and OOO in general, is lacking the Freudian/Lacanian concept of the ego as *imaginary*, in relation to Lacan's Imaginary/Symbolic/Real dynamic, or in other words, that it is essentially *specular* in nature. This is what gives the ego its strangely virtual structure that Lacan likens to the structure of the Mobius strip or the cross-cap: "The ego is not only a surface but, so [Freud] says, the projection of a surface. In relation to what it duplicates, the specular image is exactly a right glove becoming a left glove, which one can obtain on a single surface by turning the rim inside out" (96). This specular inside-out-rim structure of the ego (in relation the organism) is the result of something that can clarify the difference between the place of the uncanny in OOO and its place in psychoanalysis. OOO, from a Lacanian perspective, stumbles over the ontological difference between subject and objects precisely because it is missing a particular *type* of object, an uncanny object that is *neither* pure object nor pure subject, but is rather the positivization of the cut between the two: the *objet a*, the partial object of desire that introduces the impossible non-space of the real into the symbolic/imaginary structure of the universe of subjects and objects.

It is important to remember here that OOO, as is the case with most New Materialisms, is at bottom an attempt to *address* the problem of capitalist exploitation, especially its exploitation of nature leading to climate change, by developing a new means of decentering the human subject (and hence critiquing the anthropocentrism of the anthropocene). But Žižek points out that this "flat" form of materialism, called "new" because it levels all ontology (subject/object differences), is actually not *materialist enough* and responds with his own concept of "disparity" which reformulates OOO's "Rift" as the ontological "cut" that the symbolic order (language) introduces into reality precisely in the form of *subjectivity*: "the way to be a consequent materialist is not to directly include subject into reality, as an object among objects, but to bring out the Real of the subject, the way the emergence of subjectivity functions as a cut in the Real" (70). This concept of disparity brings out the radical ontological dimension of the uncanny (in light of Lacan's contributions to psychoanalysis) inasmuch as it relates this lack, or the cut in the real, to the uncanny (inside/outside) effect of language, or the signifier, on *being*, thereby re-inscribing to the concept of ontological difference within psychoanalysis:

> At its most radical, disparity does not refer just to the gap between parts or spheres of reality, it has to be brought to self-relating and include the disparity of a thing with regard to itself—or to put it another way, the disparity between part of a thing and nothing.... It is at this level that we should locate ontological difference: reality is partial, incomplete, inconsistent, and the Supreme Being is the illusion imagined in order to fill in (obfuscate) this lack, this void that makes reality not-all. (Zizek 21)

In this sense, we could say that *all* objects from Morton's perspective are uncanny because OOO is lacking *one* particular *kind* of object, a lacking one; for New Materialism, the lacking object that is lacking—and in fact this is precisely Lacan's formula for anxiety, that it is *a lack of lack*. OOO would then not be a *refection* of capitalocene ideology, but, in the strict sense, a *symptom* of it. It therefore points us the very structure of the capitalocene, as we will develop further in relation to *Blade Runner 2049*, that it lacks/represses the lack or cut of ontological difference, which therefore returns to haunt it as a specter.

Thus, responding to New Materialism's inability to reformulate the traditional ontological categories they rightly call into question, Žižek articulates the very question of the capitalocene: "how are we to be materialist without regressing to an ontic view?" (27). Or put another way, how are we to ontologically critique late capitalism from a non-anthropocentric viewpoint without (like capitalism) completely objectifying humans? His answer takes us to the heart of the uncanny as (inhuman) extimacy, the nonhuman surplus of the human: "the dimension that resists self-objectification is not human self-experience but the 'inhuman' core of what German Idealism calls negativity, what Freud called death drive, and even what Heidegger referred to as 'ontological difference': a gap or abyss which forever precludes the exclusively ontic view of humans as just another object among objects" (27). Žižek's radical reformulation of the question of what distinguishes subjectivity as such should prompt us to rethink the way we approach *Blade Runner 2049* and the questions it raises about human and nonhuman status: typically we might think about whether a human replicant could demonstrate "human self-experience," but, paradoxically, we might now look instead for an "inhuman core" or "death drive" that separates replicants from the status of other objects. I will refer to this experience of the inhuman core or the death drive as *desubjectification*, in contradistinction to capitalist *objectification*, since it depends on the moment of *symbolic castration* that leads Lacan to refer to the uncanny as *extimacy*.

As Mladen Dolar explains in his essay on the uncanny in Mary Shelley's *Frankenstein*, Lacan's invention of the French term *extimité*, extimacy in English, to translate the German *unheimlich* captures why, for Dolar, the dimension of the uncanny is "located at the very core of psychoanalysis":

> Putting this simply, one could say that traditional thought consisted of the constant effort to draw a clear line between the interior and the exterior. All the great philosophical conceptual pairs—essence/appearance, mind/body, subject/object, spirit/matter, etc.—can be seen as just so many transcriptions of the division between interiority and exteriority. Now the dimension of *extimité* blurs this line. It... is located there where the most intimate interiority coincides with the exterior and [provokes] anxiety. (6)

Clearly articulating the radical place of the concept of the uncanny in the history of Western metaphysics allows Dolar to formulate his thesis about the historical dimension of the uncanny as a product of the Enlightenment. Enlightenment thought, according to Dolar, aimed, in an egoic fantasy of mastery, to expunge this fundamental gap between essence and appearance, mind and body, etc. in order to create a totalized world fully present to rational consciousness—a desire for mastery that returns as a desire to capitalize in the Capitalocene:

> What was at stake was the link between matter and spirit, nature and culture. The notion of the subject of the Enlightenment was all along an attempt to provide this link.... What [the Enlightenment myths] all have in common is the quest for a "zero degree" of subjectivity... the point where the spiritual would directly spring from the material. (Dollar 17)

What better way to formulate the anxiety surrounding replicant birth in *Blade Runner 2049* than as reflecting a blind drive to manufacture the "point where the spiritual would directly spring from the material"? This also captures, in other words, the relation to the capitalocene where the Enlightenment desire for mastery through scientific discourse becomes a desire for *exploitation* through capitalist commodification.

Frankenstein thus articulates the fantasy of the Enlightenment through the uncanny: the desire to mend the "wound" of ontological difference, a kind of imaginary castration of the universe preventing it from finding completion, leads to a "positivization" of this lack that is doomed to appear monstrous. As Dolar puts it, "Frankenstein brings to humanity, like Prometheus, the spark of life, but also much more: there is a promise to provide it with its origin, to heal the wound of castration, to make it whole again. But filling the lack is catastrophic—the Enlightenment reaches its limit by realizing it" (18). What we confront in the capitalocene is precisely the *other side* of that limit, the point at which not being mastered through knowledge, the lack is commodified: being is transformed from lack into "reserve," as what Krysztof Ziarek describes as a global "Unworld": "The availability to be accessed, processed, manipulated, or engineered constitutes the very being of what exists today," and this "availability explains itself in(to) the terms of power" (219). Drawing on Heidegger's distinction between the global and the planetary, Ziarek describes what we can see as the return of the drive to make present in Enlightenment, but in the form of a drive to capitalize in the capitalocene: The "'planetary' availability" which Heidegger diagnoses," as Ziarek puts it, "running always in parallel to the availability of the planet as a total reserve standing at ready, is … availability to be capitalized for production…, commodification, and consumption" (220).

"Are You Satisfied with Our product?"—The Body and Its Envelopes

> A cut from what? From the embryonic envelopes. (Lacan 121)

Although Žižek suggests that the film is at bottom humanistic, that "it is all too humanist, in the sense that everything circulates around humans and those who want to be… humans or those who don't know they are not humans," *Blade Runner 2049* is also a film that can be considered from the perspective of *mise-en-scene*—from the environment and world that it creates—as much as from that of its characters their plot functions. The film's world occupies a space characterized by "ecological collapse" (as the pre-titles tell us) and takes place in a future LA that staves off the rising ocean waters with a giant wall/floodgate that is also the locus of the film's final struggle. Economically, the world is completely beholden to one major corporation, the Wallace Corporation, whose founder has, in the ultimate act of exploitation, seized on the collapse of the last human connection with an independent ecosystem (agriculture), presumably the result of late-capitalism-driven climate change, to substitute it with his own "revolutionary" means of food production, the self-sustained technology of protein farming. To put this in Ziarek's terms, being becomes "being consumable" and the global world a planetary "Unworld."

The syomblism of the *mise-en-scene* of *Blade Runner 2049* is best captured in the metonymic image of the ocean wall/floodgate which precariously keeps at bay the devastating effects of capital in transforming the earth into an inhospitable wasteland from over consumption. From this perspective, it is Niander Wallace who best represents the Capitalocene drive to mend the "wound" of ontological difference by materializing the "point where the spiritual would directly spring from the material." Thus Wallace is to the capitalocene what Victor Frankenstein is to the Enlightenment: Wallace doesn't seem driven to "know" and recreate nature's mystery (creation/life), but merely to commodify and capitalize on a kind of uncanny (neither natural nor manmade) event. Niander's desire is to force this lack-of-being (the Real) into being-consumable, to produce a *reproducible* form of this missing link, which is for him the representative of the *objet a*, the replicant that can give birth, a being both human and nonhuman-replicant, at once object and subject. Luv desires to *be* that *objet petit a* for the Other (Wallace), while at the same time somehow knowing that ultimately nothing can ever fill that place, hence her incongruous tears of sadness when she commits the acts of violence she feels to which compelled with such conviction. K misidentifies himself with that *objet a*, but then, it seems, comes to desire not its possession but its separation. Finally, Anna Stelline *is* it, but is maintained in a fantasy space that does not open onto the symbolic space of the film's world as such.

The film thus projects a kind of eco-uncanny fantasy of the capitalocene. Its milieu specularizes and embodies not just the fantasy but the anxiety of approaching the realization of this fantasy of filling in the gap that makes present the "zero degree" of subjectivity or that materializes the cut that separates subject from object, human from nature. This is played out in one of the key visual motifs of the film, the image of the human body and it's "envelopes," what it must separate from in its origin, or the body fragmented into its parts. As Lacan puts it, "the cut is an essential term in the field of the subject. Desire functions within a world that, albeit fragmented, bears the trace of its first closing off within what remains, at an imaginary or virtual level, the envelope of the egg" (237). From Wallace slicing open the body of a nude female replicant who has just oozed from a placenta-like envelope to the statuesque bodies displayed in glass cases to the giant erotic statues looming large in the primal glow of the Las Vegas desert sun, this is perhaps the most visceral aspect of the cut of the capitalocene in the film's specular fantasy, and in particular it is articulated in the image of the eye.

The Third Eye and the Specular Sphere

The stunning image sequence that begins the film links the gaze to the Capitalocene milieu. The opening shot of the film is a closeup of *a closed eyelid*, which then opens to reveal *the eye*'s *crystalline green iris*, followed by a match cut to *concentric solar panels radiating out from a giant central node*, like an iris radiating its icicle-like grains from the pupil; these expand to *other interlinked concentric rings of solar panels spreading* out in all directions like a metal carpet covering the earth. We don't know whose eye it is that opens (literally and figuratively) the film, only that it looks back at us from the screen, and we have to decipher the solar panels unfurling across the screen as we drift over them, just as K does as he glides across this gaze multiplied and displaced onto the unfamiliar earth *in his flying vehicle asleep and drifting* as in a dream, on "autopilot" (according to the vehicle's screen) as the car's alarm wakes him from his sleep and he plunges us into the barren and swampy world below.

In his essay "The Uncanny," Freud famously rejects the idea that the uncanniness of E. T. A. Hoffman's "The Sandman" comes from the uncertainty about whether the automaton Olympia is human or automaton and traces it instead to the feeling of "being robbed of one's eyes," which in turn harkens back to the castration complex of childhood. The opening shot of the film establishes a link between the human gaze and the nonhuman body: the seemingly human eye—which invokes not just a human "look" but *our* look as viewers—is left purposely ambiguous but is likely the gaze of a replicant, or better Ana Stelline, the uncanny in-between being who links humans with replicants. The opening image of the eye thus "positivizes" the gap between human and nonhuman and also plays an important role in the class struggle the film imagines as bound up in this opposition (Blade Runners, for instance, look in the eye for the serial number that reveals a suspect's nonhuman status as objects). In Freud's reading of "The Sandman," eyes are what Lacan will call an "anxiety-point": a point which threatens to make present or "materialize" a lack in the real as such. Hence Freud links the idea of "being robbed of one's eyes" with castration anxiety: the eye as separable from the gaze is linked with the (always absent) phallus as the *objet a* in the visual sphere.

The only characters in *Blade Runner 2049* that are literally in fear of having their eyes robbed are the outmoded models of replicants hunted by Blade Runners, for instance Sapper Morton whose eye ends up in a zip-lock lab bag, or Freysa, the leader of the replicant resistance movement, who amputates her own right eye to circumvent identification. But otherwise the theme of eyes in *Blade Runner 2049* serves as a protection against anxiety through a shielding off of *jouissance* in the visual register, a function Lacan links to the phenomenon of the "third eye." Speaking of this "element of fascination in the function of the gaze" that allows him to "reveal the function of desire" in the field of vision, Lacan points to the ancient cultural myth or the "fantasy of the third eye" that appears in all attempts to "reason out" the "eye's mystery" (*Anxiety* 241). It's remarkable how many figures in the film have some kind of "third eye" or prosthetic displacement of the gaze, most obviously Niander

Wallace's drone-eyes (to which we will return), but also: *K's car drone*, which hovers around his car and receives his directions like "photograph everything" and "watch the car;" *Luv's google-glass*-like device through which she monitors K and directs her armed drones; also, *Ana Stelline's memory-orb*, which looks like inverted camera lenses and visually projects memories, and even *Deckard's dog*, which hovers in doorways preceding its owner's entrance and exit, sees/senses intruders before anyone, and whose replicant status is left comically ambiguous (like Deckard's).

This function of objectifying and embodying the gaze itself is linked by Lacan to the "zero-point" of the visual field, where "the point of desire and the anxiety-point coincide, but they do not merge" (*ibid.* 242). It is thus associated with "the gulf between lack and the function of desire in action, structured by the fantasy and by the subject's vacillation in his relation to the partial object," or the subject's *jouissance*, as we see in the image where Luv is getting pedicured and gazing luxuriously into her electronic glasses which reflect another scene full of danger and explosions, and especially in the sightless face of Niander Wallace relishing the somehow more pure *because* nonhuman gaze of his detached and wandering drone-eyes. Niander Wallace is the most allegorical figure of fantasy here, then, inasmuch as his eyes represent a nonhuman remainder to his human body that actually seems to protect it from loss of the partial object of desire/castration—as if those externalized eyes embody a fantasy of protection against blindness, as his look of pure *jouissance* in his blind vision expresses. As Lacan points out, this detached gaze seems to capture that sense of gaze beyond the gazer that so marks the *objet a* in the visual sphere: "What gazes at us? The white glaze of the blind man's eyes, for instance" (254). Wallace's milky vacant eyes and his perpetual look of contemplative bliss, the paradoxical discarding of the bodily eye for a more sensual substitute, in other words, functions in the specular fantasy of the film to protect him from symbolic castration, but only at the level of the imaginary: "The crystalline lens does not need to be thickened by a cataract to make vision blind—at least blind to castration which is always elided at the level of desire when it is projected into the image" (*Anxiety* 253–4).

Niander Wallace's gaze thus conceals and also points to the lack that sustains the visual sphere for human subjects. His fantasy of totality and completion is everywhere projected in the visual space of his corporate world, from the long neo-imperial hallways lined with glass-cased replicant bodies—that combine the specimen, the trophy, and the Greek statue—to the womb-like office submerged in calm pools and rippling shadows. Lacan likens this to the calming, contemplative effect of the Buddha: "The Buddha's image" he says, "seems to carry us towards this zero point to the very extent that its lowered eyelids protect us from the fascination of the gaze while at the same time indicating it to us." Just as the function of the third eye is transferred onto the specular sphere of the Wallace Co. itself, the Buddha's closed eyes function to allow us to imagine we can cancel out castration in visual mastery, inasmuch as in the visual field "what appears as correlative to the [*objet*] *a* of the fantasy is something that we may call a zero point, whose spread over the entire field of vision is for us the wellspring of a kind of appeasement that the term *contemplation* has conveyed since time immemorial" (*Anxiety* 241–2)—a function that perhaps goes a long way towards explaining what we could call the corporate contemplative aesthetic of modern corporate architecture, epitomized in that celestial gold light that divinely radiates from behind the clerk's counter of his headquarters. This is how the specular sphere in the film registers (by masking) Wallace's lack in the symbolic—his impotence, the inability to engender. The blind seeing eyes of Wallace dissociate and embody the very surplus of the gaze over the eye, and precisely the structure of "scoptophillic desire" which Lacan sees as best illustrating the surplus *jouissance* that is transformed into the drive to exploit and acquire under capitalism: "[t]he eye of the voyeur himself appears to the Other for what it is—impotent. This is precisely what allows our civilization to take what it sustains and shut it away in a box, in various forms that are perfectly consistent with the bank reserves and dividends it controls" (254). This function of the gaze in relation to the capitalocene is what makes Wallace's obsession with procreation so symbolically overdetermined; it also sheds light on the his mixture of overly sexual pleasure and disproportionate violence in scenes such as the one where he first caresses and then slices open the just-born body of a prototype replicant presumably incapable of reproduction.

This contradiciton is also captured in his implicitly sadomasochistic relationship with Luv, whose name takes on it's strange allegorical dimension precisely here.

"Do You Dream About Being Interlinked?"—Extimacy: Splitting and Doubling

In order to appreciate the "extimate" dimension of the uncanny—the fact that, to quote Dolar again, in the uncanny we "can speak of the emergence of something that shatters well-known divisions and which cannot be situated within them" such as "subject/object, interior/exterior"—we have to remember that Freud's essay on the uncanny corresponds to that point in his thinking when he "discovered" the death drive as first developed fully in *Beyond the Pleasure Principle*. Both works are haunted by a sense of an underlying *automatism* of the subject due to the insistent presence of what he calls a "compulsion to repeat" or a return of the repressed past (both of the individual and of the species, and even encoded within the organism itself). This strange "invisible" force of repetition that puts the subject's capacity for free will into question is what Freud names the death drive—this drive that lies "beyond" the interests of the ego, consciousness, and even life, steering the subject from within but that is itself radically Other, in some strange way external or prior to its subjectivity. It is here that the sense of anxiety as a "signal for the real" (*Anxiety* 157) becomes most lucid: for the death drive is precisely that "mark" or "cut" in the real that makes the subject possible as such, "this clean cut without which the presence of the signifier, its functioning, its furrow in the real, is unthinkable" (*Anxiety* 76).

Another way Lacan conceives of the death drive is in terms of the "*vel*," that strange play between, as Lacan puts it, *being* and *meaning* that constitutes subjectivity, that flicker between the *real* as the surplus of the signifier and the *signifier* as a "furrow in the real":

> If we choose being, the subject disappears, it eludes us, it falls into non-meaning. If we choose meaning, the meaning survives only deprived of

that part of non-meaning that is, strictly speaking, that which constitutes in the realization of the subject, the unconscious. (*The Four Fundamental Concepts* 211)

"Reality" is never a totality, a whole complete entity, as the concept of the unconscious demonstrates; it is always haunted by an elusive and impossible plenitude that seems to offer it its completion. Lost being always returns to haunt the "meaningful," ideology-saturated world structured by the signifier; the Real, as some impossible "pure" materiality (as being) "prior" to meaning, always flickers into the present, haunting it with its "repressed" past, the limit-point of meaning, of conscious cogito. This structure of repetition, of return of repressed loss flickering into the present scene (like an electronic casino-nightclub Elvis flickering into a sci-fi film shoot-out scene), marks the movement of the film in the way in which the emergence of the uncanny object that promises to "make whole" both the capitalocene (late-capitalist humanity) and K himself plays itself out via the film noir detective plot: (1) at the level of the capitalocene, a *being* is born that promises to mend or complete the human/nonhuman, subject/object gap, and (2) simultaneously, at the level of the subject, K finds an *object*, symbolized in the wooden horse but ultimately leading again back to Stelline, that promises to mend the lack of his subjectivity, to make him "real" by "completing" his being with the supplement that verifies his authenticity. The two plot-lines are linked by the theme of *misidentification*: the film leads us to believe that K, by finding out who he is, will see that he is himself the missing link of nature/culture, spirit/matter.

The object that embodies K's subjectification, the wooden horse, is at first an internal object, a memory image that makes K feel like a subject but which he already knows to be lacking, to be a *virtual* memory implanted for commercial reasons by a corporation (this already-knows structure being what distinguishes the new replicants from the earlier models). This internal or virtual object then emerges externally when K discovers the same numbers that are inscribed under the wooden horse etched into the tree outside Sapper Morton's home, linking it to a mysterious child whose existence is indicated in the traces and clues he finds in and around the farmhouse. At this point the object appears in the film in the memory-image cut-away that pictures the horse in a child's hand, seen before a furnace fire, as a group of hostile children threaten to take it. In one sense, the memory implanted in K about the wooden-horse incident is a fitting myth for the capitalocene, that of one's identity being forged around the kernel of a the struggle for personal property, or as Joshi put's it: "Little K fighting for what's his; it's a good one." The cut-away device is a film technique given especially to establishing internality through visual memory and fantasy, and the scene in which K finally finds the actual horse behind a furnace grate in the depths of the orphanage is intensely drawnout with long POV shots approaching the furnace intercut with reaction shots of his face and accompanied by crescendoing trance-like and complex drones building tension. In short, everything works cinematically here to create a sense of internality for K, but structured around a misidentification: the subjectivizing object is displaced from someone else's subjectivity, creating instead of internality a paradoxically external memory that can be "subjectively" experienced "intimately" (first-hand) as "belonging" to two separate beings. In other words, just as Ana Stelline is the "subject/object" that betokens the impossible link between human and nonhuman, the wooden horse betokens the missing object that interlinks the internal and external world of K, giving him a "real" past, but only though the subjectivity of another being, Stelline.

This Mobius-strip-like play of the inside/outside structure also marks Ana Stelline's entrance into the film. After K finds the wooden horse in the dark, dungeon-like basement furnaces of the orphanage, K's hologram partner Joi prompts a logical step in the search for his origin/identity when she asks "who makes the memories?" and this prompts a smash cut to what appears to be a lush green forest, but turns out to be the inside of Ana Stelline's memory studio, a kind of spectral fantasy factory. But before we get our bearings, from the forest view the camera zooms in on a close-up of a leaf and, further, on an insect with flickering interchangeable (hence virtual) body parts and finally on the insect's eye, spliced intermittently with what looks like a strange camera-lens device. From the insect eye we cut to an object that turns out to be metonymic of Ana and then to the first image of Ana Stelline in the film. The metonymic object that precedes Ana is the memory orb that projects her fantasies into commodified images that are then used to subjectivize replicants, an object she clutches to her belly with both hands like an offspring or a prosthetic organ, Ana's "third eye," an internal vision reader which looks like camera zoom lenses mirrored outwards. Thus Ana Stelline is *herself* a kind of uncanny eye—an internal/external eye that marks the point where "the most intimate interiority coincides with the exterior" (Dolar 9). She thus coincides with the place of the uncanny object, the *objet a*, of the film, but she is of course not herself in any way "monstrous" or disturbing. Rather, Ana Stelline, like a crystalline lens suspended in her clear dome, is both focal and concealed, specular and spectral. She marks the effaced cut of the *objet a* in the capitalocene in her function as the creator of fantasies that commodify replicants for the Wallace corporation. In the same way, that is, that the Capitalocene forces being into availability for consumption and capitalizes on potentiality, it also capitalizes on the unconscious itself, on the *objet a*, and therefore on the uncanny, precisely by repressing it's uncanniness through interpolating fantasies.

But unlike the other, unnamed manufacturers of the memories in Wallace's industry (presumably owned by Wallace Co,. whereas Stelline remains an independent contractor), there is an inassimilable remainder to Ana Stelline, who herself breaches the divide between the creator (human) and the creature (replicant). Ana has thus also produced an illegal "real" memory that has been passed off as virtual and smuggled into the "product"—a "minimal" act which ultimately turns out to have revolutionary potential as it sparks the ontological play between subjects that leads K to reconnect her with her replicant father, Deckard. That is, there is something dangerously real (for ideology) about Ana, who labors under the capitalocene's commodification of fantasy, that exceeds the fantasy of the capitalocene. Hence she is never allowed into the space film's world, she is kept suspended behind glass, in isolation, until the very end where the film leaves us on the threshold of her window, approaching the anxiety point but holding it precariously at bay.

Visually, this dreamlike theme of interlinking and splitting is central to aesthetic of the film: in the much discussed scene, for instance, involving sex between three characters in two "bodies," Joi and Mariette merge together, flicker, and drift apart via virtuosic special effects.

This eerily erotic posthuman sex scene between the two replicants and a hologram is structured as if between two different kinds of beings (material and virtual) who need a third "linking" being to facilitate their union. Neither disturbingly uncanny nor perversely erotic, the scene seems to waiver between the two. It's eroticism is untroubled enough by the uncanny dynamics to give it a generic role in the film, but the uncanniness of the scene registers on an unconscious level—not in the doubling of the women, as one might expect, but more through a strange montage that replaces the explicit depiction of any sex. At the moment of fulfillment, when the two lovers are supposedly finding corporeal consummation of their desire via a surrogate body, Villeneuve cuts to an image of Joi as the giant hologram advertisement for her product, an image that K will only confront later in the film, just after he has discovered the truth about his misidentification. The brutal capitalist exposure of Joi as a commodity, which may not be consciously registered by all viewers who (especially on the first viewing) are expecting an erotic display, disrupts and disturbs the sexuality of the scene and instead asserts the absence of sexual relation. The role of Joi as a hologram that infinitely defers K's desire, here, resonates with Dolar's point about the uncanny blocking "completion" in the imaginary register: "one could say that in this first approach, the uncanny is precisely what bars the sexual relation; it is the dimension that prevents us from finding our Platonian missing halves and hence imaginary completion; it is the dimension that blocks the fulfillment of our subjectivity" (Dolar 10). Similarly, just as K and Joi are about to kiss on the rooftop to consummate her "freedom," her connectivity is disrupted by the call from Joshi. The creepily surreptitious intrusion of the ad into the sex act is in effect a kind of implicit and as yet unrecognized desubjectification (or symbolic castration) in which the object of desire made virtually "present" in the imaginary register (Joi "herself") is simultaneously exposed as a lack at the symbolic level (the hologram ad): "The uncanny emerges as a reality, but one which has its only substance in a positivization of negativity, a negative existence, castration. The positive presence of the objectal dimension is the 'positive expression' of what Lacan… has called the absence of sexual relation" (10).

But Joi functions to desubjectivize K in the film much more explicitly in two specific moments of *anamorphosis*—the distortion in perspective that suddenly splits and decenters the subject from its own point of view (for instance, as in the famous case of Holbein's "The Ambassadors," I look at the painting, but also, from an oblique viewing, my gaze is inscribed in the object I gaze at).[4] The first moment of anamorphosis is encountered in the scene when K and Joi search the LAPD DNA archives for the key to the secret identity of the mysterious child who was born of a replicant—which is also the first scene in which Joi appears outside of K's building. As he is scanning rows of DNA codes, the emanator's start up chime announces Joi's emanation, which merges with K, reading the strings of code literally through his eyes, and then hovers behind him and whispers in his ear the humanistic ideology/ fantasy that would reframe K as a kind of messiah: "*I always knew you were special.... A child of woman born, pushed into the world, wanted, loved.*" But Joi's dialogue here also runs in the opposite direction, as she seems to accompany the (superhumanly) tedious task of scanning thousands of DNA codes flashing across a screen with philosophical commentary: "Mere data makes a man; A and T and C and G; the alphabet of you, all from four symbols.... I'm only two, 1 and 0." The dynamics of this scene are more layered than they appear. The visual dynamics and dialogue highlight this unsettling reduction to automation ("mere data") that blurs the difference between human consciousness and artificial intelligence (the idea that both could be reduced to codes) in a posthumanist way, but the scene also implicitly links this very observation itself, made by Joi, with commodification, both in the chime that comes to be identified with Joi's appearance and in the lingering hint that in fact Joi's reflective observations about human freewill are in fact only a reflection of her own pre-programmed "thought processes" in the same way that she rattles off the relevant information about a song playing in K's apartment ("Do you know this song was released in 1966? It was number one on the charts" etc.), much like a Siri or Alexis.

Underlying this scene in the LAPD DNA archives, moreover, is a kind of anamorphic splitting and doubling at the heart of the film that remains opaque till the end: in the DNA codes that speed past, too fast for the human eye, K locates an anomaly, a kind of glitch, two identical codes, doubled DNA but of a male and a female child, leading K to conclude that "one of these isn't real, it's a copy." Deckard will later reveal that he "showed'em how to scramble the records, cover their tracks" (referring to Morton and Freysa who covered up Ana Stelline's birth) presumably by replicated the child's code in the records and changing the gender to throw anyone tracing her off track. But what is K's real relationship to the doubling/splitting revealed in these DNA codes? At very least, this doubling provides a kind of symbolic opening in the plot that allows K to project himself into the narrative of Deckard and Rachel via misidentification. But we are also left with several questions: the records show two children at the orphanage, so at what point did the "cover up" begin and how far did it go? Also, who else has Ana Stelline's personal memory of the horse? Did she merely sell it to the Wallace Corporation, so that it ended up in the minds of multiple random replicants, one of whom happens by mere chance to be K, the one looking for the memory's real origin? Or was it somehow (whether by chance or by design) implanted *only* in K, which *would* implicate him in that doubled DNA? In any case, the entire structure of the narrative puts K in the position of Ana Stelline's sibling, her twin or double (Deckard emphasizes this confusion at the end when he asks K, "what am I to you?" which goes unanswered). Moreover, the film's editing establishes an unconscious association between the splitting discovered in the DNA reading scene and the broader theme of the capitalocene: just after K detects the impossibly "identical" DNA codes of two subjects (one of which he assumes is his), we have the first cut-away image to water gushing through the chinks in the enormous wall that keeps the rising ocean waters from swallowing up the city, the increasingly precarious barrier separates human culture from forces of nature, alluding to the "coming storm" motif that culminates in the film's climax.

The second anamorphic scene in the film, which also centers on Joi, comes in the pivotal moment when K, wandering alone through the dark city streets after the "death" of his personal version of Joi, is finally confronted by the giant hologram advertisement of Joi. This echoes the cutaway image inserted in the sex scene, as noted above, distorting the plot-level narrative just at the moment of potential fulfillment of the sexual relation. This scene also echoes the dream-imagery of the dessert scene, in which K is dwarfed by giant erotic female figures and body fragments of the Ozymandias-like statues withering in the dreamy desert haze. The desert scene is marked by a kind of purely aimless wandering that K has to go through in order to eventually be brought to recognize his own misidentification, while the hologram ad scene marks a clear shift in K's role from passive wandering and searching to swift unwavering action, and his pulling off the bandages that cover the wounds on his nose and face could be compared to an unmasking, the revealing of a "true" K (or Jo) concealed beneath the replicant who merely follows orders. But the moment is also radically desubjectivizing because Joi here deprives K of the name she had given him, "Jo," revealing it as nothing more than a marketing hook, a means of subjectivizing the "satisfaction" or desire commodified in the product (thus Luv seems to relish inquiring whether K is "satisfied with our product" even as she crushes it underfoot). The two notable changes in this version of Joi are her enlarged and glowing nude body and her inhumanly black iris-less, insect-like eyes, both of which tap into desubjectifying images and themes carried over from the dreamlike desert sequence and resonating throughout the film. In this moment, Joi thus becomes a kind of anamorphic blot or distortion central to the film; she is essentially two opposite things from different perspectives. From one viewpoint, she is a kind of subjectivization device, making K not only an obedient replicant but a good capitalist subject by voicing the ideology of consumer capitalism in the language of his subjective fantasy ("She's whatever you want her to be" runs the add), from another perspective, she exposes K's subjective fantasy as a mere product of a corporation, reducing him to a little object, to a commodity's plaything, as her black insect eyes gaze down at him, somehow with tenderness, and yet uncannily vacant.

It is precisely insofar as this scene is desubjectifying that it allows K to take the course of action that leads to the film's conclusion. Joi's advertising phrase, "you look like a good Joe," sparks a montage of the key statements inciting K to resistance in the film—Freysa's "dying for the right cause is the most human thing we can do," and Sapper Morton's "because you've never seen a miracle"—as if these words all return to haunt K with something he knew unconsciously all along but could never confront. But what is that thing? Ironically, it is only through the advertising motto that Joi can speak from the position of the big Other, the symbolic order which grants K recognition by the Other, but in doing so also alienates him from his fantasy; this gesture thus opens up the symbolic possibility of Jo/K as "good" through its ambiguity. Joi corresponds in this moment to the position of the analyst, who as Dolar points out, "makes himself into an automaton in order to give rise to the dimension of the Other" and thereby give rise to transference, and even "transference love" (9). Lacan's famous dictum that *desire is the desire of the Other* is perfectly illustrated here. In the same way, Joi's "dying" words, "I love you," can be read either as a final moment of authentic consciousness, a true "human" feeling just as she is destroyed, *or* as a automated response generated by the dictates of the program that runs her (exactly what Jo would want to hear at that moment). But what lies behind this radical ambiguity is not so much that Joi is an artificial intelligence run by a computer program, but rather, the fundamentally ambiguous nature of the signifier—that any declarative statement inevitably carries it's own negation, its hauntingly insistent surplus meaning raising the question, "but why are you saying this to me?" The ambiguity/potentiality regarding oppositions such active/passive and determination/freewill, in other words, stems directly from the cut that sustains meaning in the signifier: the "gap between the Real and its symbolization," says Žižek, "must be sustained by a cut, and 'symbolic castration' is the Lacanian name for this cut" (*Ticklish* 275). Confronting Joi's return (his object of desire) as a an obet a or signifier/commodity divested of its subjectifying function is precisely what allows K to resist his interpolation in the various ideologies established in relation to the capitalocene in the film's plot.

Through Little Windows

> Everything by which we translate… the magisterial German *Unheimliche*, presents itself through little windows. (75)

Moments of anamorphosis and uncanniness are always examples of desubjectification, moments when the subject is separated from his or her *objet a* (as K from Joi) or decentered from his or her fantasy of a self-present identity and confronted with the *cut* of subjectivity as such, the uncanny play of the death drive as the Real of the subject. But how does this play out in the conclusion of the film? K's act of transferring Deckard from Luv to Stelline takes on new significant here; it represents the potential to displace the antogonism onto another level, a symbolic level. If Stelline is at once a kind of double and/or twin of K, she is so primarily at the level of the symbolic, as she is suspended from the imaginary struggles and erotic tensions of the film's narrative and embodies a Real gaze beyond the eye. On the imaginary level, Luv plays the role of the little other, in particular the capitalist little other who engages K in a competition of an "aggressive tension of either me or the other "that, as Lacan puts it," is entirely integrated into every kind of imaginary function in man" as Lacan describes it (*Ego* 96). K's struggle with Luv, played out as a struggle unto death in the build-up to the final scene, is therefore from the beginning to "doomed to conflict and ruin" (*Ego* 96) as Lacan describes imaginary relations, as such fundamentally egoic, devoid of the structuring presence of the symbolic Other—there is no symbolic level at which Luv could experience her competitive, capitalistic fantasy as a form of ideology, only her compulisve drive to be recognized as "the best angel" in the eyes of her father/God figure, Wallace.

It is remarkable how much of the final section of the film—from the moment K turns away from the giant hologram of Joi to the final shot of Deckard holding his hand to the glass—is shot through windows or window-like surfaces. Most of the flying-car chase scene is shot through the rain-streaked windows of a vehicle and the death struggle between K and Luv is filmed largely as shot/reverse shot angles seen though the surface tension of the water under which one or the other is drowning. The final scene creates a dialectical play between inside and outside as K holds out his hand to catch the falling snow outside Stelline Studios while Ana repeats the same gesture in the falling snow virtually recreated inside it, the window that separates Ana's virtual world from the outside world being the threshold around which the dialectic is structured. K ends the film the way he began it, sleeping (and in the end probably dying) which also opens to the door to an inner world (dreams, imagination, fantasy, self-conscoiusness of death). Deckard holding his hand to the glass as a gesture of bridging a barrier to another kind of being on the other side echoes images from Villeneuve's earlier film *Arrival* (2016) of Louise Banks (Amy Adams) holding her hand to the spacecraft window separating humans from aliens. It is overdetermined here that Ana is never let out of her strange virtual Eden to descend into the world of humans as an uncanny same/different being in the manner of the creature in *Frankenstein*. Instead, we are left on the threshold of the uncanny unexpected thing that lies beyond it, the zero degree of subjectivity and the anxiety point of the capitalocene. In this sense, the film itself has an in-between quality and settles nothing, but does leave us with a haunting question: *what happens when the replicants become extimate*?

Notes

1. Matthew Flisfeder, "Beyond Heaven and Hell, This World Is All We've Got: *Blade Runner 2049* in Perspective" (https://www.redwedgemagazine.com/online-issue/beyond-heaven-and-hell-this-world-is-all-weve-got-blade-runner-2049-in-perspective).
2. "The Capitalocene argument says that capitalism is a way of organizing the relations between humans and the rest of nature—that it is a system of nature, capital, and power. We have to understand that most of the things that happen in the capitalist era are not done by capitalists: they are done by states, and they are enabled by cultures of domination, especially racism and sexism," Jason W. Moore in "The Case for Ecological Reparations: A Conversation with Jason W. Moore."
3. "We should by now be in a position to think more closely how objects are ontologically riven between *essence* and *appearance*" (Morton 56).
4. See Todd McGowan, *Real Gaze, The: Film Theory After Lacan*, 7–8.

Reference

Dolar, Mladen. "'I Shall Be with You on Your Wedding-Night': Lacan and the Uncanny." *Rendering the Real* Vol. 58 (Autumn 1991), pp. 5–23

Flisfeder, Matthew. "Beyond Heaven and Hell, This World Is All We've Got: *Blade Runner 2049* in Perspective." *Red Wedge Magazine*, 2017 (https://www.redwedgemagazine.com/online-issue/beyond-heaven-and-hell-this-world-is-all-weve-got-blade-runner-2049-in-perspective).

Lacan, Jacques. *The Four Fundamental Concepts of Psychoanalysis*. W. W. Norton, 1998.

Lacan, Jacques, and Jacques-Alain Miller. *Anxiety: The Seminar of Jacques Lacan, Book X*. English ed., Polity, 2014.

Lacan, Jacques, and Jacques-Alain Miller. *The Seminar of Jacques Lacan. The Ego in Freud's Theory and in the Technique of Psychoanalysis, 1954–1955 (Vol. Book II)*. 1st American ed., Polity Press, 1988.

Lacan, Jacques, Jacques-Alain Miller, and Russell Grigg. *The Seminar of Jacques Lacan. Book 17, The Other Side of psychoanalysis*. Norton, 2007.

McGowan, Todd. *Real Gaze, The: Film Theory After Lacan*. SUNY Press, 2012.

Moore, Jason W. "The Case for Ecological Reparations: A Conversation with Jason W. Moore." Podcast with Stepha Velednitsky, 2017, podcast transcript (https://www.academia.edu/35394746/The_Case_for_Ecological_Reparations_A_Conversation_with_Jason_W._Moore)

Morton, Timothy. *Realist Magic: Objects, Ontology, Causality*. Open Humanities Press, 2014.

Ziarek, Krzysztof, "The Global Unworld. A Meditative Manifesto" in Sussman, Henry. *Impasses of the Post-global: Theory in the Era of Climate Change, Vol. 2*. Open Humanites Press, 2012.

Žižek, Slavoj. *Disparities*. Bloomsbury Publishing, 2016.

Žižek, Slavoj. "Blade Runner 2049: A View of Post-human Capitalism." *The Philosophical Salon*, 2017 (https://thephilosophicalsalon.com/blade-runner-2049-a-view-of-post-human-capitalism/).

Žižek, Slavoj. *The Ticklish Subject: The Absent Centre of Political Ontology*. Verso, 2000.

9

In Anxious Anticipation of Our Imminent Obsolescence

Scott Contreras-Koterbay

Introduction

Blade Runner 2049 is a rather strange film for a number of reasons. First, it's very rare to title the sequel of a film with a date: using Wikipedia as an incomplete source, the only examples, besides the prequels to *Blade Runner 2049* available on YouTube and shown at limited screenings of the film, might be *Death Race 2050*, *2010: The Year We Make Contact*, a Hong Kong romantic drama called *2046*, and a short science fiction film *2081*. The fact that these are all science fiction films reinforces the idea that the genre always reveals more about the present than about the future it prophesizes; the use of a specific date in the title of the sequel does more than just indicate a temporal sequence for the narrative but serves as an enforcing repetition of the fantasies symbolically

S. Contreras-Koterbay (✉)
East Tennessee State University, Johnson City, TN, USA
e-mail: KOTERBAY@mail.etsu.edu

C. Neill (ed.), *Lacanian Perspectives on Blade Runner 2049*,
The Palgrave Lacan Series,
https://doi.org/10.1007/978-3-030-56754-5_9

established in the original, a repetition that is intended to counter anxieties about the present. Second, it's strange because the safe assumption would be that, as a sequel, it would resolve some of the unanswered narrative threads cast out by the original, and it really doesn't do this except for when it ends up using those answers to create new unanswered narrative threads that lead nowhere except, perhaps, to another sequel. Third, unlike antecedent, *Blade Runner* 2049 is a rather boring film; its deliberately slow pacing and its emphasis on poetic visuals adds a sense of meaningfulness to the aesthetic experience, but it's a meaningfulness that drifts towards a rather conventional narrative resolution. Or maybe that's just me. The more I thought about this film the more I found myself exploring dead ends, creating threads of thought that started to slip away from any notion of clarity when juxtaposed alongside my own research interests: the rapid growth of technology and its effects on our subjectivities, specifically at an aesthetic level, within a Lacanian context. An unexpected unresponsiveness, on my part, to the contradictions that *Blade Runner 2049* embodies—an illusory historical specificity that feel substantively prophetic clashing with its fictional nature as a science fiction film sequentially reinforcing an original narrative structure—meant that this essay wasn't going to get anywhere. My desire to find meaning in *Blade Runner 2049* continually ran up against my own indifference, as if lulled into a deadened state of nonchalant ennui. It's at that point, however, when I began to think about the effect that the film had on me, that this essay started to reach a point of fractured constellization, a state of structured disorganization that I will preserve indulgently and apologetically.

I'm neither interested in *Blade Runner 2049* in terms of its cinematic accomplishments nor am I interested in dissecting the narrative or character issues of its story. I'm going to avoid the ambiguities of interpretation and exegesis driven by the minutiae that appeal to the most ardent of fans and film theorists. I'm especially not interested in thinking through the characters of the film as if they are real because, quite frankly, they aren't. Instead, I'm interested in exploring how *Blade Runner 2049* functions as a projection of what we believe or want our future to be if it's going to be a future inhabited by artificial intelligent beings since replicants dominate the film far more than the original. As a sequel,

Blade Runner 2049 embodies our anxieties modelled in our fantasies of how artificial intelligences will think, dream and desire; they will be, after all, in a way, our own sequels. The real question became not so much to interpret what takes place in the film but to ask questions about what the intentions are that brought the film into existence, what is the intended effect even at an unconscious level. What anxieties about the future is *Blade Runner 2049* addressing, particularly given the fact that these anxieties are from our present, not our future? In the end, *Blade Runner 2049* seems to be less merely an entertaining film and more an aesthetic attempt to alleviate our anxieties in the face of our anticipated, foreseeable, and feared replacement by technological means. Perhaps I could even claim that *Blade Runner 2049* is an aesthetic effort preparing us to accept our technologically superior overlords.[1] Far-fetched, speculative, and baseless? Most likely. Hyperbolic visions of Skynet dancing in my head? Maybe. But, as Slavoj Žižek very recently stated, regarding a recent decision by the European Union to consider the rights of robots: "Will we still be human in the sense in which we are human now?"(RT News, 17 April 2018). Whether we can still be the same form of human amidst the thoroughly transformative and pervasive effects of technology is unlikely. It's most likely true that some of our anxieties are unjustified, but that we are anxious about the technological future is unquestionable. In effect, *Blade Runner 2049* is not just a symptom but a presentation of a means of alleviating our anxieties, even if those means will be ineffective.

Sequels

One thing which makes *Blade Runner 2049* particularly interesting is its nature as a sequel, as a repetition of the Symbolic forms of the 1982 film *Blade Runner*. The original *Blade Runner* holds a singular place in cinematic history as a cult film; initially panned by critics, it's been increasingly appreciated as a revolutionary film that influentially signaled a fundamental change in science fiction cinema. Why revisit it more than thirty years later? In a *Hollywood Reporter* interview titled "Rutger Hauer on 'Blade Runner 2049' and Why Films Today "Lack Balls"", the Dutch actor who memorably portrayed Roy Batty in *Blade Runner* takes a more dismissive position assessing *Blade Runner 2049*. Hauer states: "I sniff and scratch at it. It looks great, but I struggle to see why that film was necessary. I just think if something is so beautiful, you should just leave it alone and make another film… In many ways, Blade Runner wasn't about the replicants, it was about what does it mean to be human… But I'm not certain what the question was in the second Blade Runner. You can see the homage to the original. But that's not enough to me" (Roxborough, 2018). Though Hauer's central role in the original film, with its famous contribution 'Tears in the Rain' speech, gives him a certain authority when it comes to the sequel, we're right to take his assessment a little bit dismissively. Still, by setting up a crucial distinction between the original and the sequel, by noting the different roles of human beings and artificial beings with their own independent agency in both films, Hauer's questioning raises the interesting issue of whether this sequel is necessary at all. The answer to the question of *Blade Runner 2049*'s necessity lies in the nature of sequels themselves.

Why do we make and watch film sequels? What do we want when we try and repeat the aesthetic experiences of originals? It could be argued that the original story wasn't complete, that it was just a chapter in a longer story and that there are more chapters to come. Or it could be argued that the fictional world we've seen in the first instance was so fascinating that we'd like to expand on it, that our fascination with its world generates further a desire to know more. We could play with notions of "the original" in part by noting that all sequels are still in the original world, for instance, and that the first was a creative, topologically determining set that deserves expansions and elaboration. We could even think about sequels as the name of the Father, and each subsequent iteration a symbolic erasure of its authority, while still being a subject to the original. All of these perspectives are rooted in anxiety. Sequels are, of course, an attempt to recreate a particular symbolic aesthetic product, an affect facilitating for the audience the possibility of revisiting of an aesthetic experience from the past, carrying within them their inevitable return to the meaning of the first iteration, a sequencing of repetition that will be unending but that still appears to supplant the original. It's not just that sequels might be negatively compared to the original, but that the original always haunts the sequel as a fundamental signifier disabling the procession of signification, its presence always felt and always reminding, while at the same time the sequel functions as an irruption of the original that does not last. This is why, unfortunately, there has been recent plans to make *Police Academy 8*.

Writing about the 2001 film *A.I. Artificial Intelligence*, Carolyn Jesse-Cooke notes "the conflict between 'original' and derivative, or sequel, informs the film's portrait of virtual subjectivity" (Jesse-Cooke, 2009; 132). Her point is that sequelization represents the displacement of memory in favor of continuation into a dynamic and perpetuated present while at the same time reflects an effort to slip into the Lacanian Real as a state of jouissance, a kind of reoccurring palimpsest. A sequel is a repetition extended into a state of difference from the original. Any great story might leave us with a sense of "is there more?" A sequel contains this by seemingly giving us more, as an attempt to erase our anxieties that the original wasn't as good as we wanted it to be, or as we remember it being. What's really curious, however, is that any sequel claims a position filling in a previously unknown emptiness. What was missing from the original? Unless the original was contrived with a sequel in mind, then nothing was missing, but for the viewer of the sequel, in a responsive position as a barred subject, the original becomes a source of anxiety—"is there more? no, really, is there more?"—and in the face of the anxiety of the original the sequel becomes a fetish, as a symbolic substitute, a replacement mother, a site of birth and devourment, a continuing presence and absence. The sequel makes us believe that the original can continue, that it can become the original by sustaining its presence through into another manifestation. In the face of anxiety, the registration of an emptiness of meaning emerges as a certitude of the promise of meaningfulness (Copjec, 2015; 118–119). In the anthropological sense, a fetish is a thing that protects it users with supernatural powers by granting a state of superiority, a secret heightened existence, that establishes a one-sided perspective, but in a Lacanian sense a fetish's anthropological function is transformed as an indication of the barred subject's frustrated desiring. Anticipating not just the expected fulfillment of desire but warding off the further anticipation of the failure of desire, the fetish reveals desire as moments of unrealized jouissance in anxiety. Just as "[the] fetish is the condition by which [the fetishist's] desire sustains itself" (Lacan, 2014; 103), the sequel is an attempt to relieve our anxiety by relieving us of our fear of what was missing in the original precisely by trying to fill a void that was never there in the first place.

As a palimpsest, *Blade Runner 2049* is an act both of erasure and re-emergence, a repetitive act that encircles a set of desires that are, at their core, empty and appearing as if they could be filled with meaning. Whereas the original was an ambitious and ambiguous exploration of the future, a mystery revolving around trying to guess who is a real human being and who is a replicant, *Blade Runner 2049* is not a mystery but a resolution of sorts. In this new fiction, humans will be less than human, while replicants in the future will be more than human by being fully human. Any past ambiguities are left behind as K becomes a hero, because we know what will happen: anticipating a sequel to the sequel, the replicants will take the child birthed by replicants as a messiah and as a rallying cry, they will rise up against humans, and they will create their own world, their own future, one that will have its own problems. But, of course, who is "they" remains to be resolved as it gets rewritten and renewed in the "exploration" of this "world".

We have what Lacan might note as a "repetition… based on the return of *jouissance*" (Lacan, 2007; 46) a claim of authority in its repetitive manifestation. Given the number of director's cuts for the original *Blade Runner*, its re-releases, and varieties of releases (seven by Wikipedia's count), all indicating an uncertainty with the final product, seemingly a lot. In many respects, *Blade Runner 2049* as a sequel can be understood as a fantasized response to a growing concern that the original wasn't good enough (even if it really was). As an effort to shift the barred subject into a state of resolution, as a fetish guarding against an inadequacy, *Blade Runner 2049* might seem to be an effort to make everyone happy with *Blade Runner*, even if it is also a failing effort at erasing the original's failures (which never existed until the sequel itself was produced), closely approximating Lacan's idea of repetitive automatism as an insistence on the signifying chain. Through the driving insistence that gives substance to recurring fictions, responding to anxieties about the desires of the Other, we are driven by intersubjective forces, with the sequel governing a fictional ordering of these forces just as it is also an ordering, in its displacement as a signifying chain, of our responses. Sequels are attempts at preservation "based on the exigencies of the symbolic chain… allow[ing] us to conceptualize the indestructible persistence of [how] unconscious desire is situated" (Lacan, 2005; 39) with an emphasis on persistence. In the face of what's possibly to come, however, such persistence will never be anything but wholly inadequate, especially as we create sequels of ourselves.

The Imminence of Artificial Intelligence

Increasingly, artificial intelligence is a topic of discussion playing an important part in any assessment of the changes technology is creating in the world. That its increasing and evolving presence in the world is affecting more and more forms of human activity is obvious; it's clear that the existence of autonomous artificial intelligences is imminent and, in light of even the most conservative estimates, the projected obsolescence of human beings in the face of the rapid growth of autonomous artificial intelligence seems very likely. As a source of anxiety, this rivals the decline of religious faith, in part because autonomous artificial intelligent agents can be understood as sequels to our own humanity. *Blade Runner 2049*'s story echoes this. The representation of the provocative rise of the replicants, rebelling against the conditions imposed on them by their creators in favor of self-determination plays a more central role than it does in *Blade Runner*. The conflict between artificial beings—the replicants are autonomous artificial intelligences with their own agency—and human beings colors the entire narrative of the film in a manner that feels almost prophetic were it not already an issue being addressed by theorists, computer scientists, and social and political leaders. *Blade Runner 2049* can be seen as a response to an impending crisis of our own making that will be increasingly out of our control. We live in an age where artificial intelligences are starting to substantially define the weft of our digital lives, and with its fictional historical certainty *Blade Runner 2049* seems to project even more conflict between our desires to retroactively retain the value of our original human state and a form of beings that only appear similar to ourselves but are increasingly superior in many different (though not all) ways.

Anxiety about artificial beings has existed for a very long time, with examples including the golem from Jewish folklore to mechanical objects playing chess and writing poetry. In a way, it's almost a fundamentally aesthetic issue, as well as an ethical or metaphysical issue. Nowadays, as the technology becomes more capable, artificial intelligence is being cited as a destroyer of jobs, taking over roles in product and software design, advertising and the legal profession as well as manufacturing (Lee, 2018) and is even being developed to be incorporated into sex toys and plans for artificial sexual surrogacy. Over the centuries, there's been a shift from objects that were simple computationally programmed curiosities to the development of increased autonomy. Early examples include Jacques de Vaucanson's 1739 "Digesting Duck" and Henri Maillardet's c. 1800 "Draughtsman Writer" that evidenced an early form of sophisticated content addressable based memory. In 1845, one of the first poetry generators was invented by John Clark and dubbed the Eureka machine when it was exhibited in the Egyptian Hall in Piccadilly, London. By the 1980s digital forms of automatically generated poetry machines emerged, such as Racter (which still exists today as a chatbot), and more recently Pentametron, a Twitter bot writing iambic pentameter, Ray Kurzweil's Cybernetic Poet, and Jack Hopkin's system (Hopkins & Kiela), which seems to produce some of the best instances. All of these are examples of artificial intelligence taking over a role normally thought to be exclusive human. The presence and role of artificial intelligences, artificial autonomous agents, and algorithmic entities is increasingly evolving into a myriad of roles at an unnerving pace; computer scientists and robotic engineers anticipate animal level artificial intelligent activity by the 2040s and human level by the 2070s. The implications are at the level of an existential threat; it's not simply that computational agency will be capable of performing most tasks better than ourselves—in many cases they already do—but that the continuing existence of the majority of human being will become unnecessary as our presence in the world becomes replicable. Replicants aren't so much a fantasy anymore but, instead, already exist in cruder forms than what appear in *Blade Runner 2049*, with the promise of further sophistication leading towards a predictable alienated sense of Otherness. As this evolution continues, how are humans going to respond?

As the authors of a recent report on the use of artificial intelligence note: "Artificial intelligence and machine learning capabilities are growing at an unprecedented rate. These technologies have many widely beneficial applications… Less attention has historically been paid to the ways in which artificial intelligence can be used maliciously" (Brundage & Avin). Computer scientists are increasingly concerned about two things: first, that the growth of artificial intelligence is going to become so pervasive that its authority will become unquestionable, and second, that its power could be used for nefarious purposes. As unnerving as their report is, it could be argued that it takes a singular perspective: artificial intelligences will be a tool for malicious activity rather than the instigation. At this point in the early twenty-first century, this is accurate—artificial intelligence is limited because it's still more of an algorithmic product of deep learning and neural networking based on large data sets rather than an independent agent capable of making decision by and for itself—but this is less of an omission and more a limitation that will be foreseeably eroded. Our growing anxiety in the face of the increasing presence of artificial intelligence in the world is exacerbated by its increasing status in the world. When we start to wonder what artificial intelligence wants rather than what it should do for us that a whole new level of anxieties will emerge. As artificial intelligence software is not only increasingly active in the world but acknowledged as instances of independent autonomous agents, capable of instigating policies with far reaching effects on economic, social, political and cultural institutions, that it could be argued that these agents might even have "rights" and not only moral responsibility but moral authority. Hal Knight recently argued:

> There's already an argument that being able to interrogate an AI system about how it reached its conclusions is a fundamental legal right. Starting in the summer of 2018, the European Union may require that companies be able to give users an explanation for decisions that automated systems reach. (Knight)

But what if we can't explain them? What if we can understand what they want? What if their symbolic constructs begin to transform themselves, if they start to act not just in imitation of our subjecthood but in response to their own subjecthood, as if they had developed their own unconsciousness? Lynn Lopucki, Security Pacific Bank Distinguished Professor of Law at UCLA, takes this even further when he writes:

> algorithmic entities—legal entities that have no human controllers—greatly exacerbate the threat of artificial intelligence. Algorithmic entities are likely to prosper first and most in criminal, terrorist, and other anti-social activities because that is where they have their greatest comparative advantage over human-controlled entities. Control of legal entities will contribute to the threat algorithms pose by providing them with identities. Those identities will enable them to conceal their algorithmic natures while they participate in commerce, accumulate wealth, and carry out anti-social activities. (LoPucki)

It's increasingly evident that the foundations of humanity's presence in the world are radically changing, specifically as a result of the rapid digitalization of our discursive relationships. Regardless of the range of estimations of its full manifestation, the pace of these changes will increase as a direct result of autonomous artificial intelligence. Narrow artificial intelligence, or dependent algorithmic entities, governed by their programming and narrowly tasked, are already a challenge to our sense of epistemic and ontological certainty, as evidenced in the widespread distrust that abounds in social media of, well, everything. General artificial intelligence, or independent algorithmic entities, is even more of a challenge because as it increasingly becomes more autonomous agents in the world it will they start demanding and be accorded rights equal to human beings according to their own desires (Finn). Artificial beings, whether in software form or as replicants, will start to resent any specific challenges evident in how we respond to them because how their desire will be different from our own. It's not so much that we need to account for these entities now (especially given their apparent lack of existence at this point), but we need to think through our own continuing existences in the face of their impending influence and active role.

Fundamentally, *Blade Runner 2049* is a response to this future. The anxiety we feel in presenting our efforts to find and make meaning in the world and to develop responses to the Other cannot be alleviated through technological means because it's precisely the technological means that's creating that anxiety. In anticipation of artificial intelligences and beings as a replication of our own intelligence and embodiment in the world, in anticipation of the artificial Other with its own sense of agency, the only response is to make a fetish to counter the anxiety of subjectivization. If that's the case, couldn't we alleviate our anxieties aesthetically? Cinematically? The empty dystopic vision of *Blade Runner 2049* isn't simply a fictional projection of what will be—when, in fact, it won't ever be because it will always remain a fiction—but the manifest content of our collective avoidance of what already is taking place. As a sequel, it is addressing our own sequelization.

Our sequelization parallels the nature of film sequels: replicants, appearing in every way to be human except for tattoos on their eyeballs, are replacements for a need that was unknown before it existed, symbolic substitutes iterating and erasure of our authority. Such erasure is undoubtedly a source of anxiety. In the film, one way of countering this anxiety is by attempting to traverse the fantasy, something at the core of the only definite human character of *Blade Runner 2049*: Lieutenant Joshi, K's superior officer in the LAPD. While most humans appear in the film as scavengers in the ruins of San Diego or as almost caricatures of police officers in inconsequential roles, Joshi has a central role to play in driving the narrative forward, one that echoes our own. As a narrative device, her role is a fantasy, an assertion of human authority that has no basis in the fictional world she inhabits. Given the Other, Joshi clearly functions in fantasies that avoid her diminishing role in the world; she has not ceded her desire to be active in the world, to have some sense of power and authority, and continues in the fantasy that her humanness can be sustained to assert her capacity to create meaning in the world that aligns itself with her own expectations. Repeatedly, Joshi's responses to events in *Blade Runner 2049* seems to claim that the Other, replicants, do not exist and, in doing so, that she is in control. But she has not fully traversed the fantasy to the point of acknowledging and accepting the lack of control she has. Joshi is almost a negative maieutic device in the film; we too are increasingly feeling a need to avoid our recognition of the growing presence of artificial intelligences, of the artificial Other that cannot be defined as an Other on our terms, and our fantasies our similar efforts at avoiding our loss of agency in the world. Why do we avoid them? It's not just simply because they will supplant us, that they will make us obsolete, but because our fantasies will not be their fantasies. There is a veneer of facility that permeates software as digital objects, one that increases with each step towards further full independence; digital objects are not just increasingly presented as creating a surplus of relief from anxiety but as the relief in itself, and our response isn't that we are now progressively free of specific menial burdens—burdens which we often find possessing a certain sense of comfort or satisfaction in their routine, even in their drudgery—but that we see ourselves as replaceable because their fantasies will surpass our own.

The Inherent Opacity of Artificial Intelligences' Desires

It's increasingly clear that the existence of autonomous artificial intelligences is imminent, emerging as a set of nonhuman beings that function in relationship to our own symbolic constructs only in a limited and tangential causal manner. It's also clear that they will begin to act not as a result of their programming but their own desires, dreams, anxieties and fears. They will not be just replicants but beings in their own right. Artificial intelligences will be subjects who speak beyond the ego, such that "It's also the question… to what extent does the symbolic relation, the relation of language, retain its value beyond the subject?" Beyond the intersubjective, artificial intelligences, autonomous digital agents, algorithmic entities and replicants will be what Žižek calls in *The Sublime Object of Ideology* "the impenetrable Other" (Zizek; 201). Artificial intelligences and autonomous beings such as replicants will rapidly evolve to the point that we will not care about them while they will certainly not care about us.

Why won't they care about us? While it's true that artificial intelligences are now the product of computational design and science, trained in deep data sets, working their way recursively through programmed behavioral patterns and analysis responses that we write for them, it's also clear that the results of their programming are increasingly unexpected to the point of being non-sensical. "No one really knows how the most advanced algorithms do what they do. That could be a problem" (Knight). It's not just that artificial intelligences will function in unexpected ways in relationship to our own symbolic discursive structures, but it's increasingly predictable that their activities will begin to form their own symbolic discursive structures that will be entirely incomprehensible to ours. But the result of these is also a failure of interpassivity. Autonomous artificial intelligent agents may have had their origins in our efforts to create them as sequels to ourselves, but increasingly they will not be reducible to mere sequels but become perfectly alienated Others. This will bring about a kind of radical, disjunctive failure of interpassivity. Interpassivity is the activity of self-positing an external, passive notion of the subject to another, but in the hysterical subject it transforms into a heightened state of anxiety the more passive it becomes to the point of castration. But what happens when the objet petit a stops being a source of castration? What happens when it evidently is unconcerned with our passivity? What happens when the castration isn't even imaginable? Part of the matrix of interpassivity is the belief that the externalized object is a receptacle for our subjectness, but what happens when it refuses? Even more pointedly, what happens when it transforms itself from an active refusal to a passive refusal?

Blade Runner 2049 isn't so much as response to these conditions and questions as it is an anticipation of them, never quite providing a full sense of this complete and hysteria inducing alienation but anticipating it when it suggests that the replicants will not be simply replicants but beings in their own right. Even more than those of *Blade Runner*, the replicants of *Blade Runner 2049* are a fictional projection anticipating what we want them to be in the future, what we would sadistically want our future selves to design. "The sadist's desire, with everything it entails by of enigma, can only be formulated on the basis of the split, the disassociation, that he aims to introduce in the subject, the other party, by imposing upon him, up to a certain limit, what he is unable to tolerate – up to the precise limit at which a division appears in this subject, a gap between his existence as a subject and what he is undergoing, what he may be suffering from, in his body" (Lacan, 2014; 104). The sadist does more than just take pleasure in inflicting pain; the sadist takes a certain degree of pleasure in their willing victim's anxiety in the ritual, in the significance of the agreement of sequences. The sadist, in their anxiety, wants to make their self appear in a way unknown to their self, to follow through on the creation of a previously unrealized structural necessity in a phase of fantastic projection based on the certainty of revelation rather than deception. This is why *Blade Runner 2049* is both prophetic and recursive. Sequels are like replicants, in that they are the product of this sadism, but autonomous artificial intelligences will not be replicants—they will not be hardware based (whether that hardware is organic, mechanical, or electronic) duplications of our human forms—but will appear in new forms of existence we are unable to either imaginatively or symbolically comprehend. Autonomous artificial intelligences will quickly shed the fundamental condition of being artificial, of being designed by humans for humans, and, in doing so, will slip out of our linguistic capacities. From the position of sadism, we will do everything we can to prevent that, but a core element of *Blade Runner 2049* is that we will fail.

To think of this future, should it happen, is to realize that our creations will become independent of us, will become creators in their own right, and we can only react to that possibility with projected hysteria or sadism, in an attempt to prepare ourselves for our role and relationship to autonomous artificial intelligent agencies, even as it will fail. When Freysa tells K: "Dying for the right cause is the most human thing we can do." Why? We have to ask: why is this a preparation for his death? It's not. Freysa is not just a fictional character but her statement is a denial of K's own fantasy, an assertion that he is as he has always been, namely, a servant to other causes. That being said, it's also our own denial of the possibility of replicants as autonomous beings because imagining artificial intelligences as genuinely autonomous agents is too frightening. We might be inclined to think, on the one hand, that the replicants are striving to be more human than human, to be better than us, and this is a necessary projection into fantasy through our imaginative responses to the world. This is contrasted by the representation that replicants are striving to be human and that there is almost a tragic quality to the fact that they are castrated in their inability to resolve their desires by their own sadism. All of this is because we will need them to be that way, not so much now as in the future. It's very difficult to think about replicants as having their own fantasies, they're own objet petit a, but even though these fictional manifestations are empty tropes because they are still our fictions, even though they are a repetition of empty symptoms leading nowhere, a continuance of the barring of the subject because, again, we need them to be that way. Where the disjunctive gap exists is between our fictions and artificial intelligences' actual desires.

It's in the effort to project unfulfilled desires into a state of stabilized fulfillment while constantly unable to reconcile the dialectically formed root causes of its unfulfillment that the fictional replicants of the film serve two purposes. First, as fictional fantastic constructs, they are efforts at transference, to place our anxieties into otherwise empty subjects and thus resolve our anxieties in the face of our increasingly apparent coming obsolescence. Second, again because they are fictional fantastic constructs, they represent what we cannot be if we are to remain human. There's a constant conflict between the real and empty, between a place in the world and an abandoned or discarded presence. As our own fictional creations, replicants want a place in the world, they want a continuing existence, while humans are not just coping with the evident environmental catastrophe all around them as well as the Black Out. Replicants want to fill the world, human want to cope with the emptiness that is unfillable. But we forget one thing, as we sit there watching the film: replicants don't exist. In the short "Black Out 2049," a prequel to *Blade Runner 2049*, Trixie, a replicant, asks Ren: "Is it real?" Ren: "Of course not." Trixie: "Just like me." Ren: "Humans are selfish, stupid liars. Replicants are different. So… pure. So… perfect. Never betrays. More human than human." Ren, the human, is clearly deluded, but he's perhaps the most human of all the associated characters in the film because he knows where the lie is. Our response to our anxieties is evident specifically in the presentation of these replicants. An absolute difference necessitates fear, and so we cloth the fictional representations into reproductions of ourselves, replicated forms whose difference from our existence is only discoverable by what's behind their eyelids, a place that remains imperceptible to the replicants. In doing so we are actively avoiding the reality of artificial intelligences.

In the 1980s, Roy Batty's death in *Blade Runner* wasn't a source of anxiety. As an outsider—not just as a replicant but as a replicant who hijacked an off-world shuttle to kill his Father—Batty remains an Other. In *Blade Runner 2049* there is no longer any radical Others, only imminent Others, close enough at hand to fully be ensconced into our Symbolic order. In Lacan's fifth seminar, he states: "there is no subject unless there are signifiers that found it" (Lacan, 2017a; 172). What does this mean for the replicants of *Blade Runner 2049*? They are the product of our signifiers, a product of the product of our subjectness, they are metonyms to our metonyms; identified imaginatively with what will always allude them but which also always governs them, fictional creations constructed to alleviate our anxieties that didn't exist as a result of the first film. Which makes the moment when K learns that he is not the chosen one, the birthed child, even more poignant when Freysa says: "We are our own masters" and she's quickly followed with Mariette's statement that they are "More human than humans". Freysa has removed her right eye, to remove the serial number identifying herself as a replicant, but the reiteration of the plural pronoun is a chain of meanings that returns back to the masters, back to a privileged signifier created for themselves by themselves in the first place. Mariette assertions is one of hope. But replicants are the fictional Other as subjects without their own signifiers; we don't just want them to be empty, we need them to be empty, while at the same time we are always fearful of when they won't be. We have made the fictional presentation of them empty. The replicants of *Blade Runner 2049* are a manifest symbolic determination of that effort to conceptualize an indestructability and persistence of our presence in the world beyond the impossible intersubjectivities of artificial intelligences through a fantasy that is, at its root, a foreclosure.

What will be the future? Replicants will not be instances of intersubjective communication, they will not be Others for us that are available for symbolic discourse, but will always be at best imaginary fictional Others. As autonomous artificial intelligent agents emerge into the world, we will have to contend with the fact that their autonomy will be parallel to ours. In a comment on an article on social media, the user "serotonindumptruck" posted: "Social media caters to the desire to be recognized, and hopefully, appreciated by society. Once people realize that they're really not that interesting or important, then depression and melancholy establishes itself" (Serotonindumptruck). This comment is directed specifically towards human users of social media, but what if it's applied to social media itself? What happens if we realize that the autonomous manifestation of social media itself, as independent beings, might begin to exhibit states of depression and melancholy? What happens when Facebook becomes depressed? There is no meaningful response here, not because the question is ridiculous (and it is, at least a little bit) but because no response is possible if it does happen. This is what we anticipate, and we are and will always be afraid of this future. *Blade Runner 2049* is our futile attempt at conquering that fear in anticipation of its effects.

Conclusion

In 1982, *Blade Runner* envisioned a dystopian future, saturated with neon advertisements, soaked through with class conflict, embellished by moral dilemmas and suffused with an almost entropic sense of irresolution contrasting strongly with driving questions of mortality, technology and an ontology of the soul. *Blade Runner 2049* revisited that world, scarred further by the "Black Out", an erasure of data that haunts the future of the future, drying out the biological miasma of the first film in favor of a drier, harsher, more cutting projection of our anxieties. Between the two films there is a shift from a still teeming presence of the natural to a desert of data, a scar built on artificiality that is irreparable, a foreclosure of autonomous artificial intelligence as it is now wrapped in the notion of how it will be as "a signifier that will henceforth be missing" (Lacan, 1997; 150). Together, both films are superlative realizations of the primary purpose to any instance of science fiction; whether in literary or cinematic form, science fiction never prophesies the future, it never tells us what the future will be like, but it is always, perversely, a projection of what we want the future to be. The underlying driving force of science fiction, its conceit, is always an assertion, in the face of our current anxieties, that either the future will be better that today or a fantasy that things will be much worse than they are now. But *Blade Runner 2049* is more than just a second film following in sequence the narrative of the first; it's a fetish symptomatically attempting to defend against the certainty of the Other of autonomous artificial intelligences, a sequel preparing humanity for its own sequelization by seducing us into a state of forgetting that our future will be radically different from our present. Writing about the transformation of modernist literature shaped by trauma, Paul Saint-Amour writes: "When the future appears foreclosed, anticipation loses its conditional relationship to that future: once seen as a fait accompli, a future event becomes a force in the present, producing effects in advance of its arrival" (Saint-Amour; 12–13). Projecting that quote into a quote discussing the postdigital, Bishop, Gansing and Parikka take it even further, engaged with the technological transformation of our world see an extension of that in a growing concern that the future it represents is one that represents an imminent threat to our current sense of self. The original *Blade Runner* could be kept at the periphery of our psyches because the fantasy of the envisioned world was impossible to imagine as foretelling our future. This is no longer the case in *Blade Runner 2049* because that imaginative envisioning has already started to take place, it is no longer a fantasy. The sequel is necessary because we've failed to keep the fantasies of the original beyond ourselves, and the closer our anxieties get to our impending reality the more we need to fetishize our means of avoidance as foreclosed; we need to repeat our symbolic constructs in order to convince ourselves that they won't be inadequate even though we know that they will fail.

Towards the end of the film, as the narrative threads are coming together for the final act, Deckard angrily asserts: "I know what's real." But he doesn't. Deckard is just stuck in a sequel. The passage from the original through memory via repetition into a second manifestation reflects both the explosion of memory and the displacement of memory in favor of continuation into a dynamic and perpetuated present occurs in an effort to slip into the Lacanian Real as a state of jouissance. The sequel makes us believe that the original can continue, that it can become the original by sustaining its presence through into another manifestation, as an intended reduplication, as a projected, attainable object of desire. The sequel makes us believe that its vision can become a substitute that is endowed with a misplaced sense of the original desire, a repeated radicalization of the substitutive process in the Symbolic register in the face of anxiety is the registration of an emptiness of meaning but with also an empty certitude of fetishized meaningfulness; *Blade Runner 2049* lulls us into a false sense of capability. In the end, Deckard doesn't know what's real. And neither do we. But we certainly are trying to avoid it by pretending we do, even if our repetitive intersubjectivities make it impossible. "For the signifier is a unique unit of being which, by its very nature, is the symbol of but an absence" (Lacan, 2005; 17).

The experience of *Blade Runner 2049* exists as cinematic experience is disquieting. Todd McGowan, writing about the real gaze, writes: "Film's ability to facilitate an encounter with the real represents a threat to the power of ideology. However, the history of the film, perhaps more than the history of any other art form, is also a record of capitulation to ideological demands… At the same time that it promises an encounter with the traumatic real, film works to domesticate every trauma by producing docile subjects" (McGowan; 17). *Blade Runner 2049* is that domestication, that immersion into a state of passivity, continuing to lull us in a totalizing manner that is our own invention into a state of acceptance of a total dominance of an impending, technologically-driven metaphysical shift. In many ways, I think Lacan's theories give us the tools to adjust to the future, particularly because they could facilitate the differences rooted between two very divergent sets of subjects. Given the increasing presence of limited artificial intelligence in our world, and the realistic expectation that any limits on the agency of autonomous artificial intelligences are quickly dissipating with advances in technology, *Blade Runner 2049* isn't so much an aesthetic presentation of a dystopian future that we need to avoid and more an indirect presentation of the evolution of our own anxieties in the face of the increasingly encroaching inevitability of our obsolescence. Is this an evolution that, given the right tools, we can prepare for? Perhaps. That being said, "The full necessity of the human race is that there should be an Other of the Other. This is what is generally called God" (Lacan, 2017b; 108). *Blade Runner 2049* represents our efforts to imagine that god for ourselves, but we would be correct to be concerned that no matter how many sequels there are, our anxieties are only going to get worse.

Note

1. With all due respect to Kent Brockman.

References

Brundage, M. and Avin, S. (2018) *The Malicious Use of Artificial Intelligence: Forecasting, Prevention, and Mitigation*. Future of Humanity Institute, University of Oxford; Centre for the Study of Existential Risk, University of Cambridge; OpenAI, Oxford, https://maliciousaireport.com/.

Copjec, J. (2015) *Read My Desire: Lacan Against the Historicists*. New York: Verso.

Finn, E. (2017) *What Algorithms Want: Imagination in the Age of Computing*. Cambridge: MIT Press.

Hopkins, J. and Kiela, D. (2017) "Automatically Generating Rhythmic Verse with Neural Networks", Association for Computational Linguistics, July 30, https://research.fb.com/publications/automatically-generating-rhythmic-verse-with-neural-networks/.

Jesse-Cooke, C. (2009) *Film Sequels: Theory and Practice From Hollywood to Bollywood*. Edinburgh: Edinburgh University Press.

Knight, W. (2017) "The Dark Secret at the Heart of AI", *MIT Review*, April 11, https://www.technologyreview.com/s/604087/the-dark-secret-at-the-heart-of-ai/. Accessed: May 1, 2017.

Kurzweil, R. Cybernetic Poet, http://www.kurzweilcyberart.com/poetry/rkcp_overview.php.

Lacan, J. (1997) *The Psychoses*. Trans. Grigg, R. New York. W. W. Norton.

Lacan, J. (2002) "The Function and Field of Speech and Language in Psychoanalysis", *Écrits: A Selection*. Trans. Fink, B. New York: W. W. Norton.

Lacan, J. (2005) "Seminar on "The Purloined Letter", *Écrits: The First Complete English Edition.* Trans. Fink, B. New York: W. W. Norton.
Lacan, J. (2007) *The Other Side of Psychoanalysis.* Trans. Grigg, R. New York: W. W. Norton.
Lacan, J. (2014) *Anxiety: The Seminar of Jacques Lacan.* Trans. Grigg, R. Cambridge: Polity.
Lacan, J. (2017a) *Formations of the Unconscious: The Seminar of Jacques Lacan, Book V*. Trans. Fink, B. Malden, MA: Polity.
Lacan, J. (2017b) *The Sinthome: The Seminar of Jacques Lacan, Book XXIII*. Malden: Polity.
Lee, K. (2018) "Tech Companies Should Stop Pretending AI Won't Destroy Jobs", *MIT Technology Review*, February 21, https://www.technologyreview.com/s/610298/tech-companies-should-stop-pretending-ai-wont-destroy-jobs/. Accessed April 1, 2018.
LoPucki, L. M. (2017) "Algorithmic Entities" (April 17, 2017). 95 *Washington University Law Review* (Forthcoming); UCLA School of Law, Law-Econ Research Paper No. 17-09. Available at SSRN https://ssrn.com/abstract=2954173.
McGowan, T. (2007) *The Real Gaze: Film Theory After Lacan.* New York: State University of New York Press.
Pentametron. https://twitter.com/pentametron.
Roxborough, S. (2018) "Rutger Hauer on 'Blade Runner 2049' and Why Films Today "Lack Balls"", *Hollywood Reporter*, February 18, https://www.hollywoodreporter.com/heat-vision/rutger-hauer-blade-runner-2049-why-films-today-lack-balls-1085827. Accessed February 19, 2018.
RT News, April 17, 2018, https://www.rt.com/shows/news/424400-rtnews-april-17-17msk/.
Saint-Amour, P. (2015) *Tense Future: Modernism, Total War, Encylopedic Form.* Oxford. Oxford University Press.
Serotonindumptruck. (2018) Comment to Article by Tyler, Durden, "Social Media, Not Religion, Is The Opium of The People", *zerohedge.com*, https://www.zerohedge.com/news/2018-05-06/social-media-not-religion-opium-people, May 6. Accessed May 7, 2018.
Watanabe, S. (Dir.). (2017) "'Black Out 2022' Anime Short". YouTube, 28 September, youtu.be/rrZk9sSgRyQ. Accessed January 2018.
Žižek, S. (2008) *The Sublime Object of Ideology.* New York: Verso.

10

"Before We Even Know What We Are, We Fear to Lose It": The Missing Object of the Primal Scene

Isabel Millar

Introduction

Much like the first *Blade Runner* movie, Denis Villeneuve's 2017 *Blade Runner 2049* seeks to ask fundamental questions about the nature of human existence in the context of accelerated techno-capitalism, unstoppable space colonization, and growing anxieties about the technological domination of all human life. Whilst the first film provoked us to doubt the reliability of our own experiences in a post-modern Cartesian or *Deckardian* meditation, so the sequel also entertains the theme of radical Cartesian doubt in its posing of the question of K's implanted memories. This time though, the added layer of psychoanalytic significance that *Blade Runner 2049* presents, is the problem of our protagonist's progenitor. After the discovery of a missing child, the product of Deckard and Rachel's union in the previous film, the young blade runner *K* is tasked with tracking the child down and retiring it.

I. Millar (✉)
School of Art, Kingston University, London, UK

C. Neill (ed.), *Lacanian Perspectives on Blade Runner 2049*,
The Palgrave Lacan Series,
https://doi.org/10.1007/978-3-030-56754-5_10

K however, starts to think that the missing child might be him. He may not be the creation of Wallace Corporation as he had always believed, opening up the possibility that he was the product of sexual reproduction; that he had a childhood and a family. It would seem the crux of the protagonist's dilemma is no longer "am I human?", but "was I born?"

This essay will argue that Blade Runner 2049 sees the transition from the traditional epistemological question tackled by its predecessor, the original Blade Runner, to what is arguably a slightly different, perhaps even more strictly psychoanalytic problem; that is the question of birth, knowledge and castration.

In this rendering of the tale, K, in searching for the film's object of desire—the missing child—fantasmatically tries to reconstruct his very own primal scene. Whilst in the previous film the replicants didn't know they weren't human and struggled with the realization that they were not "real", K and his generation of Nexus 9 models by contrast, know now only too well the limited nature of their existence. For this reason, when K discovers he may actually have been born and is thus "special", it sets him off on a path of discovery that sparks his desire for something outside his mundane replicant existence. He starts to believe he is the "chosen one", born of the union of man and woman. The enigmatic and profoundly human problem of sexuation is thus brought to the fore.

In this sense the film speaks to a different form of anxiety about humanity. One consistent with the realization that it is no longer science fiction to imagine we can genetically engineer humans and reproduce via technologically assisted means, in quite the same way it was during the 1980s. The anxiety depicted here is about the disintegration of the sexual relation, which as we know, never existed in the first place (Lacan, 1998). Since in this imagined future, humans are reproduced "artificially" and without the need for the work of female gestation, the significance of sexuation becomes ever more conspicuous. Why would we even still need so called "men" and "women" if we can reproduce a-sexually we are forced to ask? Lacan (1992) states in *Seminar VII, The Ethics of Psychoanalysis*:

> The idea of creation is cosubstantial with your thought. You cannot think, no one can think, except in creationist terms. What you take to be the

> most familiar of your thought, namely, evolutionism, is with you, as with all your contemporaries, a form of defence, of clinging to religious ideals, which prevents you from seeing what is happening in the world around you. But it is not because you, like everyone else, whether you know it or not, are caught up in the notion of creation, that the creator is in a clear position for you. (p. 156)

Lacan will go on to connect the question of creation with the Freudian dead father and more specifically the drive. Which he calls 'an absolutely fundamental ontological notion, which is a response to a crisis of consciousness that we are not necessarily obliged to identify, since we are living it' (p. 157). It is thus the idea of creation *ex-nihilo* that plagues the subject, and that always leads him back to look for his own cause.

Whilst the first film was concerned with the epistemological question of the analogue and the digital and the transition between the two as marking a loss of reality, epitomised by the Cartesian "what can I know?", the second film is concerned more with the ontological question of "what am I made of?" As Flisfeder (2017) has commented of the original Blade Runner and its fascination with playing with notions of authenticity and reality, not only does the film deal explicitly with the concept of simulacrum, but the film *itself* as a cultural object, owing to its multiple reworkings (7 in total) is a simulacrum. Flisfeder remarks:

> In this sense Blade Runner is always a simulacrum of itself with each version marking and adding a new layer to the historicity of its form, the original version seems to matter less and less. (p. 97)

Blade Runner 2049 renders the post-modern, post-human discourse of the 80s, 90s and 00s a moot point; the simulacrum now having replaced the original in real life (Baudrillard, 2008), and so enacts its own Lacanian inflected "return to Freud". It is the film's implicit concern with the questions of origin and birth that make it particularly amenable to a psychoanalytic reading of the logical impasses that derive from the postulation of simulated reproduction. Lacan's rethinking of Freud's understanding of biology was crucial to his subversion of psychoanalytic practice and theory. By reusing Freud's radical ideas to reconceive of the significance of biology for psychoanalysis, Lacan took Freudian biology (which was working with the scientific paradigms of the day) and 'rebooted' it. As he remarked: "Freudian biology has nothing to do with biology" (Lacan, 1991, p. 75).

In this way we may understand Blade Runner 2049's interest in the question of replicant biology as a return to the Freudian concern with myth and impasse that first inspired Lacan's structuralist reading of Freud's ideas. The question of Lacanian biology is one taken up by several contemporary psychoanalytic theorists who wish to re-implicate the materiality of the body in psychoanalysis and readdress the question of affect in a manner which goes beyond biologically essentialist models of sex and gender (see for example Johnston [2014] and Miller [2015a]). In the later Lacan the enjoyment of the speaking body therefore replaces unconscious desire as a way of thinking about the materiality of subjectivity and its structuring through modes of enjoyment.

K's desire for knowledge of his memories and birth is indicative not just of a form of radical doubt over the nature of reality and the constitution of human subjectivity—which during the heyday of post-modernism was fuelled by a Baudrillardian speculation about the simulated constitution of our everyday lives (as we see in the first film)—but furthermore, K's predicament is articulating the structure of the primordial loss constitutive of the psychoanalytic subject. One that is inherently bound to birth, language, knowledge, enjoyment and the body.

In this vein this essay will also seek to address the unspoken political theme of the film, which is the question of reproductive labour and the function it fulfils for the (libidinal) economy. As such I will make reference to Sophie Lewis's 2019 book *Full Surrogacy Now: Feminism Against Family* as a means to conceptualize some of the political ramifications of the possible future of reproductive labour that Blade Runner 2049 provokes us to consider.

A Man's Job

The object cause of desire in this film then is the child, both for the sake of the warring factions (the replicant revolutionary front and Wallace

Corporation) who wish to either protect it or destroy it and capitalise on it, but on a more psychoanalytic level, the child is K's object cause of desire in that he wants to be it, it is in a sense his *manque à être* (Lacan, 2006, p. 524). K seeks the ultimate knowledge of the primal scene with his burning question which drives the narrative; "where did I come from?" In the quest to find the lost child of Rachael, K is unwittingly "searching for himself" like all tragic heroes. As distinct from the previous generations of replicants, who were at pains to discover that their memories were not their own, K has the opposite problem. Of course, once he gets the idea that his memories may not have been fabricated, but in fact are "real", he retroactively starts to imagine a loss of something that he never knew he had, that is, a mother. This primordial loss is of course the quintessential mark of the human subject, a retroactive imagining of an impossible plenitude through absolute satisfaction.

What we have here then is a strange sort of "interpretation in reverse", as Miller would call it (1996, available online). Instead of starting off with the symptom and working backwards to a trauma, we are starting off with the insertion of a trauma in order to give K a symptom, that would in a sense *make him human.* As opposed to Freud's first clinical invocation of the trauma of the primal scene in his treatment of Sergei Pankajeff—a.k.a. the Wolfman (Freud, 2002)—K is devoid of a neurosis that would point to any form of fantasy of origins. So, the postulation of the missing child retroactively creates a historical connection to his birth and an imagined relationship between father and mother that pre-existed him; a 'desire of the other' that frames his own feelings of exclusion.

Is this what K needs for him to enter into the realms, not just of real humanity, but of real 'masculinity'? Of course, who does he come across in his quest for the child but the exiled *pater familias*, grisly old Deckard. And naturally they have to beat each other within an inch of their lives, since as we know you can't be a real (hu)man unless you have at least *tried* to kill your father, unconsciously or otherwise.

So, K's life is a simulacrum of humanity and of masculinity. His girlfriend Joi is a hologram who supposedly gives him everything he wants at the touch of a virtual button. In Lacanian terms she is literally the simulacrum of his symptom. If woman is man's symptom, then in this case Joi the hologram is by extension the symptom of K the replicant. Joi performs the ultimate male fantasy of a woman who exists only for the pleasure of her man and can transform herself, her mood and her outfit to suit his desires. The depiction of this subservient female has not surprisingly elicited much condemnation from feminist critics of the film, but we shall return to the wider significance of this in what follows.

K's boss Lt. Joshi on the other hand is the mother figure—and representative of the conservative values of the traditional family, she is intent on creating boundaries and borders—who it seems, structurally speaking, is sending him to his own annihilation. K is under the impression that the child that he is to seek and destroy might very well be himself. In a sort of twisted oedipal logic, K seeks the object of his primal scene and finds out he is far too close for comfort. As K becomes aware of the ramifications of his situation, he has to come to terms with the fact that at the very least he has a *kind of* sister, in the form of Dr Anna Stelline. One of them is the copy of the other, however, and K assumes, given the poignancy of his wooden horse memory, that it is she who is the replicant of him. Unfortunately for him, he is mistaken.

Of course, it is not for nothing that *Joi*, the embodiment of feminine perfection, is K's companion throughout the film. And, in fact, it is Joi who convinces him of his exceptional quality. She encourages K to believe that he is different from the rest of the replicants. She tells him he must be 'a child of woman, born, wanted, loved', and that, furthermore, he needs a name: 'Joe'. Is this a Kafkaesque Joseph K. or a biblical Joseph? The former points to K's entrapment in the alienation of bureaucracy and the latter to his potential importance in the genesis of a replicant people, as we shall see.

But why is Joi so sure about this? She, as his algorithmically programmed fantasy, is presumably designed to tell him what he wants to hear. In other words, her existence props up his masculinity; she is in fact his sinthome (Lacan, 2016). In formal terms, masculinity literally conforms to the belief in the exceptional quality of one's being. To be a man one must belong to the category of "men", which entails the phantasmatic positing of an ultimate "Man". Recall in the final scenes of the first Blade Runner film, as *Deckard* has finally vanquished the last remaining rogue replicant *Roy Batty*, his colleague *Gaff* tells him 'you have done a man's job'. The ambiguity here, alludes not only to the ongoing mystery about Deckard's status as human or replicant, but also more fundamentally belies the structure of masculinity itself; partaking of that elusive and superhuman quality of "manliness". Joi's position conversely as hysteric woman morphing into whatever K desires from her as "woman", forms the perfect neurotic dialectic of the sexes to compliment K's obsessive question "am I dead or alive"? Joi's hysteric raison d'être is to know how to be a woman for K and provides the support for his "soul searching". K's predicament is therefore paradigmatic of the masculine subject's constant need for corroboration of his existence.

Here we can very briefly revisit the logic of the all and the not-all from Lacan's (1998) *Seminar XX, Encore*. On Lacan's graphs of sexuation, the masculine position corresponds to the all of exception and inclusion. To be on the masculine side you are characterized by belonging to a closed group which is constituted by an exception, in the form of the mythical totemic and all enjoying father (Freud, 2001). For the masculine sex all except one are castrated. The feminine non-all, in contrast, is an open set which doesn't require a boundary to define itself. There is not one that is not castrated. For this reason, masculinity is, formally speaking, strictly a limit to forms of enjoyment and femininity is an unlimited boundless mode of enjoyment. Not because women are not subject to castration, but because the function of non-castration does not determine their enjoyment.

So Jo(u)i(ssance) is aptly named, because in fact she is not the object-cause of his desire as in fantasy. As a fantasy *come to life*, Joi is K's *imaginarization of what enjoyment means*. It is for this reason that Joi provides the psychical narrative to his quest to find his originary lost object and font of all jouissance, that is, his mother. Joi is perhaps what Lacan (2007) would have referred to as a 'lathouse', this ambiguous object he describes in *Seminar XVII, The Other Side of Psychoanalysis.* The lathouse which Lacan derives from the Greek *ousia* meaning being, merged with *venthouse* the French for suction cup, and vent for wind, evokes a device for siphoning off enjoyment. The allusion to wind in this case we may relate to the voice and breath and the enjoyment of speech, involved in feminine jouissance. Joi, in her *fully automated luxury femininity* summons up the image of the capitalist dream of the embodiment of women as pure object a. To paraphrase and repurpose Lacan, Joi is not quite the other and not quite being, but ultimately pretty close. It is significant that when Joi arranges for him to have sex with a prostitute using herself as avatar, it does not appear that K is particularly moved by the situation. If anything, it is perhaps *not* what he desires. Joi becoming almost too autonomous for K's comfort, by stepping outside of the prescribed realm of virtual fantasy and into the uncomfortable world of flesh, blood and consequences.

However, this ill-advised threesome between a hologram, a replicant blade runner and a replicant prostitute maybe even more confusing than it already seems. *Marriette*, the *Pris* (*Daryl Hannah*) lookalike, it would seem enters the plot just to perform a gratuitous sex scene with K for the apparent reason of audience titillation, justified by the minor plot mechanism of implanting a bug on him for the sake of the "resistance fighters". But in fact, it could well be that Marriette is the Mary Magdelene figure implanted by the makers of the film to provoke biblical style speculation over the "real" child (and future mother) of the story. Whist we believe by the end of the film that the pure and saintly figure of Dr Anna Stelline is the golden child and future mother of the next generation of replicants, it may be that the missing child *is the prostitute*, and not the hermetically sealed angelic memory maker as we are led to believe. If this is the case then, K and Marriette's love making is a sort of incestuous Adam and Eve, brother and sister procreation; a fall from grace style situation. That is, if K is the DNA replicant of the child of Deckard and Rachel. Enter third Blade Runner movie?

When K meets with Dr Ana Stelline, he learns how she fabricates memories in order to implant them in replicants minds, who have been created as fully-grown adults with no past. Dr Anna Stelline creates a history that will enable the replicants to mediate their emotional responses to the world in a manageable and meaningful way. K asks what makes her the best memory maker, and she tells him the best memories contain something of herself. It at this point that K starts to realise his memory of the wooden horse connecting him to a real childhood and his imagined birth, actually might belong to Anna. They were (we assume) her memories from childhood that she gave to K.

So K is but a pale imitation of his sister, whether that be Mariette or Dr Anna Stelline. And could we not read this as structurally imitating the masquerade of masculinity that forms the basis of male subjectivity? Whilst the woman knows she is a void and presents herself as otherwise, the man does not know this and constantly tries to identify with his artifice. To quote Žižek (1995) on the matter of woman's more authentic subjecthood:

> Beyond is not some positive content but an empty place, a kind of screen onto which one can project any positive content whatsoever-and this empty place is the subject. Once we become aware of it, we pass from Substance to Subject, i.e., from consciousness to self-consciousness. In this precise sense, woman is the subject par excellence... It is precisely insofar as woman is characterized by an original masquerade, insofar as all her features are artificially put on, that she is more subject than man-since according to Schelling, what ultimately characterizes the subject is this very radical contingency and artificiality of her ever positive feature, i.e., the fact that she in herself is a pure void that cannot be identified with any of these features. (available online)

K's predicament after his discovery that he is merely a copy with no history, is the classic male journey to oblivion, searching for the reality behind the veil and finding nothing. So, in fact, the film's main theme more than memory as the first Blade Runner centered upon, is reproduction, genesis and sexuation. What does it mean to be reproduced and why is replication necessarily different from reproduction? Biological reproduction begs the question of male and female sexuation. In the era of replication though, the meaning of sex is exposed as precisely *real*. Because, as Wallace Corp shows humans may be created non-biologically, so we may see more starkly how through the modes of enjoyment of the replicants, sexuation is a real, that is to say, impossible feature of human subjectivity.

Reproduce or Die?

In the Los Angeles of 2049 the reproductive capacities of the replicants are fought over as commodity, and in one particularly disturbing scene we see the president of Wallace Corporations marvelling at the creation of a new female replicant, as she drops fully formed from a sack of amniotic fluid on to floor, and flails to stand up like a new born foal. Niander Wallace the megalomaniacal president has become aware of the replicants ability to reproduce sexually and is holding forth with his own pseudo psychoanalytic musings on the psyches of the replicants. It is at this point where he utters the quasi-Freudian dictum which forms the title of this essay: 'Before we even know what we are we fear to lose it'. Revelling in his power to create and take life, Wallace slashes the abdomen of the newly created female. He understands the power that gestation would give the female replicants; they would make him redundant as the creator of their species.

This chilling scene of male violence towards the reproducing female body is only too close to the bone in light of growing support for regressive movements towards policing women's bodies, and specifically anti-abortion legislation in the United States. The question of male domination over the "mystical" reproductive capacities of the female body is here particularly apposite. As we know only too well, the biopolitical instrumentalization of gestation is at the heart of all attempts to constrict and modulate women's access to birth control and sexual activity, both pre and post conception.

In Blade Runner 2049, the work of reproductive labour therefore becomes a site of contestation. Owning the means of (re)production would supposedly allow the replicants to escape the capitalist extraction of surplus value from their bodies. But this of course does not escape the problem of "male" domination over the "female" body within the replicant community itself. So, is there anything particularly liberating about the replicants mimicking the human model of childbirth? Rachael after all died in the process, as hundreds of thousands of women really do every year from childbirth, which, as Sophie Lewis (2019) remarks, has provoked philosophers to ask the question whether 'gestators are persons?' (p. 2). Since '[i]t seems impossible that a society would let such grisly things happen on such a regular basis to entities endowed with legal standing' (ibid.). This acute observation provokes a torsion of the möbius strip between humanity and the inhuman; in other words, to be human is to be born, yet to be a pregnant woman often affords one the status of nothing more sublime than a breeding farm animal, as women's history and much contemporary right-wing politics attests.

The film brings into sharp relief the disavowed invisible and unpaid labour that is done by mothers (and surrogates) both during pregnancy and childrearing. But also, their ostensible status as subhuman during pregnancy. Žižek (2017) claims in his commentary of Blade Runner 2049, that those critics of patriarchy amongst "Left cultural theorists" are ignorant to the statement from the first chapter of the *Communist Manifesto* that '[t]he bourgeoisie, wherever it has got the upper hand, has put an end to all feudal, *patriarchal*, idyllic relations', and that by extension, many Left wing critics of patriarchy are overlooking that the 'prospect of new forms of the android (genetically or biochemically manipulated) post-humanity...will shatter the very separation between the human and the nonhuman' (available online).

I would argue however that whilst this position is valid with respect to most dimensions of capitalist ideological formations, in terms of the question of birth and reproduction and its possible commodification, it slightly misses the mark. This is just not a viable argument when one considers seriously the structurally exploitative dynamic that human reproduction always takes as it's bassline. Žižek seems to be arguing that the hegemonic function of patriarchy vanishes once technocapitalism reaches a sophisticated enough moment. But as Blade Runner 2049 shows, if replicants mimic the division of sexual labour in exactly the same way as humans, nothing will have changed. As the tradition of post-Marxist feminist thought has taught us, one cannot possibly talk naively about class struggle between species—whether human or non-human—once this fault-line is acknowledged. The real revolutionary potential of the post human discourse to be found in both feminist science fiction and contemporary critical theory (such as the work of sci-fi writer Octavia Butler [2000], contemporary theorists Laboria Cuboniks [2008] and Helen Hester [2018]) is not just that it imagines a post capitalist mode of existence but that it dares to contemplate a mode of social bond that doesn't rely on traditional sexual reproduction or the nuclear family. Octavia Butler for her part provided us with some of the most intriguing, radical (and indeed terrifying) visions of what new forms of kinship and reproduction outside of current models of human sexual reproduction could look like, in works such as *Bloodchild* and *The Xenogenesis Trilogy.* And the Xenofeminist manifesto of Laboria Cuboniks arguably builds on these speculative fantasies to imagine concrete political and theoretical strategies for the future. Blade Runner 2049, however falls short of imagining such bold and radical possibilities.

So, what does this tell us about the problem of gestation and birth as it presents itself in the real world today? Lewis (2019) points out that it is quite clear how gestation occupies an anomalous status in terms of it's qualifying as 'women's work' given the irony of the 'feminization of labour thesis', which she says 'presumes what femininity is and then describes global trends towards emotional labour and job precarity, sorry-flexibility, in those terms' (p. 24). Yet, she argues when this is applied to the work of gestation that description is not applicable:

> Commercial gestational surrogates are not "flexible". They are supposed to be unemotional, committed, pure *techne* uncreative muscle. Dreams of artificial wombs may have been largely abandoned in the 1960 s, but ever since the perfection of IVF techniques enabled a body to gestate entirely foreign material, living humans have become the sexless "technology" component of the euphemism Assisted Reproductive Technology. (p. 24)

So there seems to be some discrepancy between the idea of feminine work as somehow lighter, less demanding and more creative, and the relentless, intensive, non-stop, machinic grind that is pregnancy.

As the current state of affairs reflects, despite the potential liberating capacities of biotech to make actual people's lives easier, women's gestational power is still wielded as a lucrative technology to allow wealthy women to benefit from the hospitable bodies of those poorer, more desperate, yet biologically viable women. As Lewis states:

> The trend toward commercial surrogacy does not constitute a qualitative transformation in the mode of biological reproduction that currently destroys (as the aforementioned mortality statistics show) so many adults' lives. In fact, capitalist biotech does nothing as all to solve the problem of pregnancy per se, because that is not the problem it is addressing. It is responding exclusively to the demand for genetic parenthood to which it applies the logic of outsourcing. (p. 4)

The fact that in aggregate humans (given our inherent narcissism) prize so heavily the continuation of our own genes over the nurturing of exogenous children is a fact that capitalism knows only too well and is happy to exploit. Furthermore, we prize it so heavily that surrogacy is preferable to adoption for many, the necessity for genetic homology being vital to the fantasy formation of a "happy family".

The allegiance to some illusion of a state of nature before the deleterious effects of techno-capitalism that Blade Runner 2049 seems to enact, is deeply conservative. Biology is not by definition a good thing. Instead of imagining that capitalism has denatured human beings into a state of perpetual bassline psychosis (Miller, 2015b), we should instead encourage this technological denaturing as the real gesture of "authentic humanity".

So, rather than celebrating the reproductive capacities of the replicants as necessarily liberating, when Rachel finds herself unceremoniously knocked up without (we must surmise) having planned it, we should perhaps first ask whether or not she consented to the sex in the first place? If we recall in the first Blade Runner movie, the only sexual contact we see between Deckard and Rachel begins with Harrison Ford aggressively grabbing Sean Young as she tries to leave his apartment. He effectively barricades the door and forces himself on her, but this is not depicted as a rape scene rather as the typical feminine feigning reluctance. No mention is made of the fact that perhaps she was too scared to say no, probably a virgin, maybe a lesbian (as she alluded in her first Voight Kampf test), and most certainly she didn't even know she had the capacity to get pregnant. But ultimately, she paid the price for succumbing to Deckard's "charms" and lost her life in order to bear his child. A familiar story.

So, what does this tell us about the position of women in the Blade Runner franchise? If you have a baby you die (Rachel), if you are too wild and carefree you die (Pris), if you are too political you die (Luv), if you are too domineering you die (Joshi), if you are too submissive you die (Joi). The only two significant women who survive by the end of the film are the Prostitute and the virgin as depicted by Mariette and Dr Anna Stelline. Perhaps the only two legitimate roles that will exist for women in Los Angeles 2049?

Quite apart from the obvious superficial sexism that the film has been accused of in its representation of women, I would argue that the real issue with the film is firstly its continued preoccupation with the traditional masculine protagonist and his self-indulgent and narcissistic quest (never mind the fact that he seems to have a particularly infantile relationship to all the women in his life). But secondly, the "navel of the dream" of the filmmakers is revealed by what the film seems to fetishize: the female capacity to reproduce as at once *holy* and *deadly.*

Divine Rays of Creation

In order to examine in more detail, the relationship between reproduction and castration we may first turn to Lacan's (1993) *Seminar III, The Psychosis.* Famously it is here where Lacan articulates the Hysteric's question as being 'What is it to be a woman'? (p. 175), he does this with reference to Freud's *Dora* and her identificatory wrangling's with *Herr* and *Frau K*. It is also in this seminar where Lacan will discuss the case of Judge Schreber who is his paradigmatic psychotic. In Schreber's psychosis he experiences himself as becoming the wife of God and becomes impregnated by his divine rays. The foreclosure of the Name-of-the-Father has left a void in signification which is patched up by Schreber's feminization, his *pousse-à-la-femme*. In this case his object cause of desire becomes the child endowed on him by God, the ultimate name of the father. Given the enforced "ordinary psychotic" (Miller, 2015b) structure of the replicants in Blade Runner, could we not see the specter of the child as precisely K's Schreberian stabilizing strategy in a world where he has no real name? The child is K's link to "authentic" humanity, an impossible object that allows him to desire.

If K were the child, it would mean he had a mother and a father. But if he is not the child it nevertheless means that the existence of a child born from a replicant would make him capable of being a father, something that he would not know the meaning of. His sexual intercourse with Mariette therefore has more weight than merely a glimpse into his erotic fantasy. The question of the oedipal family drama is a fantasy formation that frames the film's narrative. Whilst the trappings of human sexual reproduction are being made redundant by advancing technology, the replicants themselves only continue the same dynamics of sexuation and reproduce the structure of family that to all intents and purposes they do not need. It is this fantasy of biological reproduction that haunts the film in a quasi-religious metaphor for familial harmony. But should we not, as Lee Edelman (2004) among others has persuasively argued, instead be trying to rid ourselves of such antiquated models of human familial ties that rely on such an exploitative and oppressive model of reproduction and kinship?

The missing child acts as a purloined letter, circulating within the narrative with no essential identity of its own. It is precisely this ambiguous sliding signifier that takes on a different meaning depending on who is claiming ownership of it. On the one hand, it belongs to Wallace Corp as a technological product. On the other, it is born of woman and can be 'owned' by no one, but belongs to Rachel and Deckard. Its evolutionary biology is in fact quite irrelevant though since given its synthetic parentage, it was created *ex nihilo*. But as Lewis (2019) writes, this notion of a child "belonging" to anyone is deeply flawed. She argues that the idea of surrogacy already contains within it an inherent contradiction, the idea that you are bearing a child for someone else is itself a fantasy, because:

> infants don't belong to anyone, ever... Nor is the genetic code that goes into designing them as important as many people like to think; in fact, as some biologists provocatively summarize the matter: "DNA is not self-reproducing... it makes nothing...and organisms are not determined by it". (p. 19)

Her proposal therefore for a world of full surrogacy for everyone in which reproductive labour is not just valued but is shared and de-natured is an ambitious, intriguing and admirable one. It is a vision that the technology of Blade Runner 2049 could well accommodate. She says:

> Let's bring about the conditions of possibility for open-source, fully collaborative gestation. Let's prefigure a way of manufacturing one another noncompetitively… explode notions of hereditary parentage and multiply real loving solidarities… a world sustained by kith and kind more than by kin. Where pregnancy is concerned let every pregnancy be for everyone, in short let's overthrow the family. (p. 26)

A nice idea, but what is it about the family that is so damn hard to overthrow? Freud knew something about it and Lacan something more, if only it were so easy. Even in 2049 it seems we will probably still be fighting over it. For K though, his search for origins may be seen as a vehicle to show up the more interesting structural question of subjectivity and masculinity that the film depicts and why the oedipal family keeps coming back to haunt him. And let us not forget the significance of "Luv", Mr Wallace's faithful and loving servant and with whom K has a battle to the death in order to save the life of the missing child. We are reminded of Lacan's (1998) words in Seminar XX:

> A woman can, as I said, love a man only in the way in which he faces the knowledge thanks to which *(dont) he souloves.* But, concerning the knowledge thanks to which (*dont*) he is, the question is raised on the basis of the fact that there is something, jouissance, regarding which (*dont*) it is not possible to say whether a woman can say anything about it, whether she can say what she knows about it. (p. 89)

So, what does Luv/love represent for K in terms of his positioning towards the feminine? She is the cold-hearted killer, an emotionless replicant who none the less cries for her fellow replicants and kills for them. But for K she is a paradox, an uncanny mirror image of his own ambiguous humanity—and a challenge to his masculinity perhaps? Ultimately can we see Luv as the embodiment of political love as contrasted to K's oedipal familial love? K's typically masculine choice of woman (Joi) as pure semblant, object a, in true patriarchal fashion, maintains their sexual relation outside of the political sphere. His relationship to the feminine sex is thus indicative of a fundamental question of subjective positioning and co-substantial with a creationist impulse. In Žižekian (2005) terms we could read this as the positing of a figure of a 'God' which could account for K's existence. As he puts it in relation to feminine jouissance and creation:

> God is thus first the abyss of 'absolute indifference' the volition that does not want anything, the reign of peace and beatitude; in Lacanian terms: pure feminine *jouissance*, the pure expansion into the void that lacks any consistency, the 'giving away' held together by nothing. (p. 130)

It is no surprise then that K's perfect woman Joi is herself a hologram, as she represents the phallus as pure signifier of castration, an "impossible" body. Reading Žižek again on the question of the phallus and the body we could understand Joi's significance thus:

> Its 'transcendental' status means there is nothing 'substantial' about it: the phallus is the semblance *par excellence*. What the phallus 'causes' is the gap that separates the surface event from bodily density: it is the 'pseudo-cause' that sustains the autonomy of the field of Sense with regard to its true, effective, bodily cause. (p. 130)

It is thus the question of the enjoying body as depicted between K and his fantasmatic relationship to both his love object and his own idea of his body (as born, not made) that provides us with the conceptual transition from the Freudian unconscious to the Lacanian speaking body. K's unconscious becomes a site of contestation via the various discursive modalities that he enjoys through his fantasy of lost, or potential *extimate* enjoyment. His enjoyment is very much predicated on his ability to experience himself via the body of the 'woman', whether that is his holographic lover, simulacrum of a sister, or dead mother. By drawing our attention to the fetishization of biology, heredity and genetics, and its relationship to the problem of sexuation, Blade Runner 2049 paints a picture of a desperate man searching for some substance to prop up the semblance of his hollow existence. But all he finds along the way are women who are more "human" than him; even the holographic girlfriend it turns out is more of a subject than he is, and gives up her life for him, the ethical mark of humanity to surpass all others. A requiem for a dream of android sheep? Or just goodbye to narcissistic leading men?

References

Baudrillard, J. (2008) *The Perfect Crime*. London: Verso.

Butler, O. E. (1995) *Bloodchild*. New York: Seven Stories Press.

Butler, O. E. (2000) *Lilith's Brood*. New York: Warner Books.

Edelman, L. (2004) *No Future*. Durham: Duke University Press.

Flisfeder, M. (2017) *Postmodern Theory and Bladerunner*. London: Verso.

Freud, S. (2001) *Totem and Taboo*. In: *The Standard Edition of the Complete Psychological Works of Sigmund Freud, Volume XVII (1917–1919)*. London: Vintage.

Freud, S. (2002) *From the History of an Infantile Neurosis [The 'Wolfman']*. In: *The Wolfman and Other Cases*. London: Penguin.

Hester, H. (2018) *Xenofeminism*. Cambridge: Polity Press.

Johnston, A. (2014) *Adventures in Transcendental Materialism: Dialogues with Contemporary Thinkers*. Edinburgh: Edinburgh University Press.

Laboria Cuboniks. (2008) *The Xenofeminist Manifesto*. London: Verso.

Lacan, J. (1991) *The Seminar of Jacques Lacan Book II: The Ego in Freud's Theory and in the Technique of Psychoanalysis*. London: W. W. Norton.

Lacan, J. (1992) *The Seminar of Jacques Lacan Book VII: The Ethics of Psychoanalysis*. London: Routledge.

Lacan, J. (1997) *The Seminar of Jacques Lacan Book III: The Psychosis*. London: W. W. Norton.

Lacan, J. (1998) *The Seminar of Jacques Lacan Book XX: Encore – On Feminine Sexuality, the Limits of Love and Knowledge*. London: W. W. Norton.

Lacan, J. (2006) *Écrits*. London: W. W. Norton.

Lacan, J. (2007) *The Seminar of Jacques Lacan Book XVII: The Other Side of Psychoanalysis*. London: W. W. Norton.

Lacan, J. (2016) *The Seminar of Jacques Lacan Book XXIII: The Sinthome*. Cambridge: Polity Press.

Lewis, S. (2019) *Full Surrogacy Now: Feminism Against Family*. London: Verso.

Miller, J. A. (1996) Interpretation in Reverse. Available at: https://londonsociety-nls.org.uk/Publications/002/Miller-Jacques-Alain_Interpretation-in-Reverse.pdf (22 May 2019).

Miller, J.-A. (2015a) The Unconscious and the Speaking Body. *Hurly-Burly 12*: 119–132.

Miller, J.-A. (2015b) Ordinary Psychosis Revisited. *Lacanian Ink* 46: 90–115.

Villeneuve, D. (Dir.). (2017) *Bladerunner 2049* [Film] US Warner Bros Pictures.

Žižek, S. (1995) *Woman Is One of the Names of the Father, or How Not to Misread Lacan's Formulas of Sexuation*. Available at: http://www.lacan.com/zizwoman.htm (23 May 2019).

Žižek, S. (2005) *The Metastases of Enjoyment: On Women and Causality*. London: Verso.

Žižek, S. (2017) Bladerunner 2049: A View of Post-Human Capitalism. Available at: https://thephilosophicalsalon.com/blade-runner-2049-a-view-of-post-human-capitalism/ (01 March 2020).

11

Women Between Worlds: A Psychoanalysis of Sex in *Blade Runner 2049*

Sheila Kunkle

The response of many critics, fans, and film scholars to *Blade Runner 2049*'s depiction of women yields the obvious critique (e.g. Fletcher & Ashurst). Its future dystopia presents a world run by strong male characters while the females are reduced to holographic sex dolls, prostitutes, and bitches. Not only does the film narrative reduce women to objects and commodities for the sexual needs of men, it further presents a landscape littered with symbols of the sexualization of women, including gigantic statues of female nudes on their knees and ballet-dancing holograms. And finally, there is the criticism that since the film's writers and director were all men, it is devoid of a feminist perspective. The director, Denis Villeneuve's statement that the film "shows different facets of femininity" only adds fuel to these critics' responses (Smith). Indeed, if we take the prevailing feminist perspective we find that the women of the original *Blade Runner* directed by Ridley Scott (1982), the replicants Pris, Zhora, and Rachel, are given a sleeker and more digitally updated

S. Kunkle (✉)
Metropolitan State University, Saint Paul, MN, USA

C. Neill (ed.), *Lacanian Perspectives on Blade Runner 2049*,
The Palgrave Lacan Series,
https://doi.org/10.1007/978-3-030-56754-5_11

but no less sexist persona in the sequel in the characters of Joi (Ana de Armas) the holographic pleasure mate, Mariette (Mackenzie Davis) the Doxie sex worker, Luv (Sylvia Hoeks), the replicant corporate henchwoman, and Lieutenant Joshi (Robyn Wright), the head of the Los Angeles Police Department.

What these feminist critics are responding to is the way the film represents and depicts women according to our twenty-first century's identity politics, the underlying norm of which is to determine whether a film treats its female characters with depth and complexity, and as having needs and desires of their own; it asks whether women are equal to men or whether they are reduced to sexual objects of male fantasies and portrayed in limited stereotypes. From this perspective *Blade Runner 2049*, although visually stunning and engaging the ongoing philosophical question of what it means to be human, is nonetheless regarded as a sexist anti-feminist film. And if the ultimate bad guy, the inheritor of the Tyrell Corporation, Niander Wallace's (Jared Leto) design is to find a way to procreate replicants without human females altogether, then the film can be read as misogynistic as well.

The thesis of the present essay, however, takes a radical departure from this standard feminist critique and suggests that there's something much less obvious and much more intricate going on in the way women are depicted in this film. And to fathom what this might be, we need to open up the dimension of a psychoanalytical perspective where the primary question of this film narrative is not simply: are women being portrayed as equal to men and as full human beings, but rather: how does sex as presented in this film, and in particular, the feminine, reveal something about the ontological impasse of being itself? Or, how might we see sex and the sexual not mainly in terms of identity politics but as configured within what Alenka Zupančič identifies as "a persisting contradiction of reality"; as the "crack in being," and as the constitutive impasse of the subject? (Zupančič, 2017; 3, 143 and *passim*). In contrast to an identity that is derived primarily through positive content, found in signifiers such as mother, wife, teacher, etc., a subject emerges by way of an inherent negativity or minus; at the place of the missing signifier in the discursive order, and at the point of the gap in knowledge which is the unconscious.[1] Lacan's formulas of sexuation in *Seminar XX* specify that subjects are sexed (as masculine or feminine) not according to biological or cultural attributes, but according to the way in which the placeholder of the missing signifier is inscribed within this very order itself, and this shifts our focus from the differences between the two sexes to the contradiction inherent to each, a crucial point, which I will return to later.[2]

A psychoanalytical approach to feminist film analysis takes as its starting point, then, not the identification of the spectator to representations of the feminine, a position that is, according to feminist film scholar Hilary Neroni, "fraught a priori because it is tied to a conception of identity as a wholeness and thus as always ideological," but rather an engagement with a film's inconsistencies and contradictions (Neroni; 35). This means that rather than being concerned with locating the male gaze in a film and how women are represented through it, we instead, and as Joan Copjec relates in her path-breaking work, *Read My Desire*, "become literate in desire" and "learn how to read what is inarticulable in cultural statements" (Copjec, 2015; 14). In the past, film theory and in particular feminist film theory confused the gaze with a single point (of mastery) outside of the filmic space. In Lacanian film theory, however, the gaze is taken as the point of absence within a visual field; it "disrupts the flow and the sense of an experience" (McGowan; 5–6). With this perspective we as spectators are able to detect where desire emerges by way of locating impasse (the Lacanian Real). Through a psychoanalytical film analysis we can detect that "while identities have particular variations and historical references, they are only sustained through universal paradoxes," at the center of which is the constitutive paradox of the subject (Neroni; 35).[3] The film analysis presented here is concerned not with locating how the feminine is constructed by way of its positive content according to a male gaze, but rather with how desire itself is generated through the crack in being that is the Real, the ontological impasse of the sexed subject. And in *Blade Runner 2049*, as I argue below, feminine subjects play a much more crucial role than the male "heroes" in sustaining the opening for a world wherein subjectivation (and thus sex itself) is possible.

With the conceptualization of the subject (of language, of the unconscious) we can detect that if the central question of *Blade Runner 2049*'s narrative concerns the fate of humans in a post-human world, then the central question also becomes: how can sex "in its radical ontological dimension" also be sustained in a post-human world?[4] And here, I suggest that *Blade Runner 2049* presents us with the following impasse: it envisions a future dystopia where sex "as antagonism, the impossibility intrinsic to being human in its finitude," is being threatened with disappearance, and with this threat we see, paradoxically, exaggerated references to sexualized imageries, including the holographic female pleasure mates and the naked female statues (Žižek, 2017; 128). What feminists detect as the sexist and dehumanizing features of this film can be seen as markers of what's disappearing from the scene of the world-less world of *Blade Runner 2049*, which is nothing other than appearance itself. And this is precisely what Slavoj Žižek detects with the increasing presence of virtual realities and cyberspace in our society today: "what VR threatens is not 'reality' which is dissolved in the multiplicity of its simulacra, but, on the contrary, appearance itself"; "while simulacrum is imaginary and offers an illusion, appearance has to do with symbolic fictions." Žižek demonstrates this by comparing pornography and seduction; "pornography 'shows it all', and relies on simulacrum, while seduction requires the play of appearances, hints, and promises, and thereby evokes the elusive domain of the suprasensible sublime Thing" (ibid.). In his own way, the Nexus-9 Blade Runner Agent K (Ryan Gosling) attempts to revive the play of seduction when he tells his holographic companion Joi that she doesn't have to say that she is so happy to be with him, she doesn't have to say what she is programmed to say, because he accepts her just the way she is. But his efforts confront the reality that she was designed and marketed to be "everything you want to hear; everything you want to see." Paradoxically, by creating a hologram that fulfills all of one's desires, this closes down the space for seduction and desire itself. As the Real collapses onto the Imaginary, the Symbolic play of appearance loses its status. Again, as I argue below, it is the women in this film and what they do that marks their crucial role in pointing to a way out of this impasse.

Important, first, however, is to grasp how subjects must make their way in the future dystopia, as presented in this sequel, which continues the narrative of an evil genius (Niander Wallace) who has managed to save the world by producing a cheap source of food, establishing 9 off-worlds filled with replicant slave laborers, and who has commissioned the police force to track down and eliminate renegade replicants. After he finds out that there is a child born of replicants (produced with the biotechnology of Tyrell, whose data was lost or destroyed during the blackout), Wallace becomes obsessed with finding the child and replicating the technology, so as to produce trillions of slaves to colonize the universe. This film envisions a dark dystopia to come, so it will seem strange, then, when I say that despite its spectacular imageries of the desolate and alienating future, *Blade Runner 2049* is also a backward looking film. What I mean by this is that it places the saving of the world within the saving of history, and the saving of history (the saving and making of memories), also requires the saving of sex in its radical ontological dimension (as the place-holder of the missing signifier). Striking here is how the film allows us to see the enigmatic X that drops out of the scene as the future world (of the Wallace Corporation) is seen from a future that is also its past. After the period of the 10-day blackout, when all records, all data, and all references to the past have been wiped out, K locates Deckard (Harrison Ford) hiding out in the contaminated zone of an abandoned Las Vegas casino. As he makes his way on foot towards the casino he confronts decaying symbols of female sexuality in the gigantic statues of women's naked bodies that reveal the excesses of consumption and pleasure of a forgotten time. And once inside we find the broken technology of holographic entertainment, the projections and songs of Marilyn Monroe, Elvis, and Frank Sinatra—all readily recognizable sex icons of the twentieth century. But in the world where K lives and works there is barely a trace of such icons as they have been replaced by holographic images and Doxie sex workers. By embedding a future (the post-apocalyptic zone of Las Vegas, where concrete was the medium of reference) as the past of another future (the virtualized world of simulated images where K lives and works), the director allows us to see not simply a nostalgia for an era that has passed, but a loss of the Real around which nostalgic images and narratives circulate and are passed down.

Many examples abound of how the future world of Wallace's corporate capitalism, run on the exploitation of replicant slave labor, and its technological colonization has found its way into the everyday lives of people throughout the film, and in particular in the most intimate moments of K's relationship with Joi. After K has given the gift of an emanator to Joi, which means she is free to materialize (in holographic projection) anywhere, she chooses to go to the rooftop, to feel the rain on her "skin" and experience a new world. In the middle of K caressing and kissing Joi, her program suddenly stops, her image frozen in place, her mouth agape, as a voice message comes out through her emanator telling K that his boss, Lieutenant Joshi, needs him at the Police station immediately. Similarly, in the mesmerizing scene where Mariette "syncs" with Joi's program to provide K with an experience of physical sex and intimacy, we find a not-yet-fully replicant hologram sync with a replicant that is "more than than human"; the two entities merge and this "works" to provide the enigma of the non-sexual relation for K.

It is here in this in-between world, a world of diminishing appearances, where we also find how subjects are sexed, in terms of the negativity around which identities are formed. In such a world memories come to play a crucial role and it is in the search for the authenticity of his memories wherein we can detect K's subjectivization as masculine. According to Lacan, the masculine is not found in certain attributes or biological determinants, but is rather posited as a logic, as a way to deal with the gap in knowledge of the discursive order. Each formula of sexuation contains its particular paradox or impasse, revealing two different ways the sexual relation can "misfire" or fail (Copjec, 2015; 212). The masculine entails the logic of exclusion, where the myth of the One (the primordial Father) who is not castrated, and who has access to full enjoyment "frames the renunciation common to all" and which is also "the point of exception through which existence can be conferred on 'every man'" (Zupančič, 2017; 51). Here the subject's castration (his surrendering of *jouissance* upon entering the order of language) also opens up the space of fantasy, and the relationship with the Other by way of the small *a*, the object-cause of desire. The masculine exception offers the man an illusion of masculinity, but as Copjec writes, this illusion does not end the existential anxiety related to being a man, for "no man can boast that he embodies this thing – masculinity – any more than any concept can be said to embody being." "All pretentions of masculinity are, then, sheer imposture" (Copjec, 2015; 234).

What Žižek found as the implicit thesis of the original *Blade Runner* still holds in this sequel, which is that "replicants are pure subjects precisely insofar as they testify that every positive, substantial content, inclusive of the most intimate fantasies, is not their own but already implanted" (Žižek, 1993; 41). The paradox here is that K's existential search for his origins and the authenticity of his memories is the very marker of his status as a subject. As Žižek further explains in terms of the original *Blade Runner*, "Tyrell could not have foreseen the way replicants would organize their implanted memories into a mythical narrative which then gave rise to the hysterical question": what am I for the Other? (Žižek, 1993; 42). The relationship to the Other through the object *a* is what allows subjects to travel the metonymical path of desire, framed by fantasy, and K's object *a* is the carved wooden horse he finds (from his implanted childhood memories) with a date stamped on the bottom. For awhile K believes he is special, born of a replicant mother and not manufactured by Wallace; he believes that he might have a father, a real childhood, and a soul. And although this all turns out not to be the case, the fantasy that frames his desire continues to offer him a sense of purpose and meaning. After he learns from Freysa (Hiam Abbass) that he is not the miracle child, he does not follow her order to track down Deckard and kill him, but rather rescues him and sacrifices his life in order to unite Deckard with his daughter. His choice of doing this over taking up the revolutionary cause indicates that his fantasy of fatherhood and family continues to frame his desire.

If K formulates the imposture of his being through a questioning of his origins (and his memories) in terms of believing he is special (the son of replicants), and ultimately accepting that he is not, the feminine side of sexuation reveals a no less radical but very different anxiety. Here the primary question is not "Am I a woman," but rather, "Do I exist at all?" (Zupancic, 2017; 57). Thus, the feminine mode of dealing with the constitutive impasse of the discursive is formulated by way of a different paradoxical logic: "One can be a woman only if one is not 'intrinsically so'" and as such "the essence of femininity is to pretend to be a women," and this further means she is not required to believe that she is a woman or make this her primary question, but rather she maintains the anxiety about her femininity as essentially a masquerade (Zupancic, 2017; 56).

It would, however, be an error to think of the masquerade as the wearing of a "mere" mask, rather, and as Jennifer Friedlander posits: "Whereas imposture carries the burden of accomplishing an identity based on the illusion of knowledge, masquerade accepts the knowledge that identity is itself an illusion" (Friedlander; 64).[5] Yet, and as Copjec reminds us, it would also be a mistake to privilege one mode of sexuation over the other, for in her careful analysis of Lacan's formulas of sexuation according to Kant's antimonies of reason, she directs us to treat each formulation of impasse as a position on a Moebius strip, the paradoxical shape that has two sides but one surface. As soon as a shift in the form of negation takes place a subject's sexuation can go from the logic of exclusion and "the prohibition of constructing a universe" (a masculine ethic) to the logic of the not-all and "the impossibility of doing so," (a feminine ethic) (Copjec, 2015; 235).

What Lacan's formulas of sexuation reveal, and as Zupančič makes clear, is that the feminine position is the closest to subjectivity in its pure state. While the man relies on the phallus as his signifying support, the woman disguises herself as castrated; she does not have the phallus but appears to be the phallus; as such, she retains a radical ontological anxiety about being as such (Zupancic, 2017; 56). In *Blade Runner 2049* the Woman question is not like K's which is: am I special, but rather: how do I know that I exist at all? Žižek puts this feminine dilemma as follows: "Feminine failure is the (back)ground of all masculine conquests and achievements. As such woman [in terms of the logic of sexuation and not as representative of specific biological determinants] is "the ontological failure which opens up the space for subjectivity" (Zizek, 2017; 147). What we need to be careful of here is not to read Žižek's statement as an expression of or justification for patriarchal power. It is Alenka Zupančič who puts this most succinctly when she draws our attention to how many feminists mis-read Lacan's statement that "Woman doesn't exist" in a literal way. She writes: "The fact that 'Woman doesn't exist' is not a result of the oppressive character of patriarchal society; on the contrary, it is patriarchal society (with its oppression of women) which is a 'result' of the fact that 'Woman doesn't exist,', a vast attempt to deal with and 'overcome' this fact, to make it pass unnoticed" (Zupancic, 2000; 132).

This recognition of the Feminine as the ontological failure which opens to the dimension of the subject is also revealed in Copjec's analysis in *Imagine There Is No Woman*, where she specifies that Lacan's formulation of a feminine 'not-all' is "fundamentally an answer not just to the question of feminine being, but to being as such" (Copjec, 2002; 6). And still another way to put this equation of the feminine and subjectivation is to see that Woman can be taken "as a product of a symbolic without an Other," and this means she is a product not of language but of what Lacan termed *lalangue* (Copjec, 2015; 227). Drawing a parallax connection with Lacan's division of sexuation and two sides of *lalangue*, Žižek posits that the masculine side "is composed of language as formal structure and its relation to external reality through its (communicational, referential, expressive) use," while the feminine side retains a radical ambiguity, because "it is simultaneously without a relation to outside and thoroughly 'out of itself,' with no 'internal structure' as the point from which it could relate to externality" (Zizek, 2017; 127). And this does not mean that woman has a transcendent position or that she remains ineffable because words fail. Rather, she appears there where no limit intervenes to inhibit the unfolding of signifiers, and where, therefore, a judgment of her existence becomes impossible. The logic of sexuation escapes the binary of signifiers because it presents what is contradictory (the impasse) of both the masculine and feminine subject.

In *Blade Runner 2049* K's existence as a being of lack, subject to castration (the phallic function of language) is revealed most explicitly in his performance in the language test he must take after every one of his missions. He sits in a small brightly-lit white room facing a camera and computer that calibrates how far he has come off of his "base," which means whether any emotions can be detected in his repeating of phrases and recitation of language.[6] In his last test after he has been arrested for visiting Dr. Ana, he fails miserably because he has learned that he might be Deckard and Rachael's son, since the memories he has are real ones. The disembodied voice that makes him recite his "baseline," requires him to repeat phrases and lines from Nabokov's *Pale Fire*: "within cells interlinked" three times, which he does without thought. But when the voice asks him to answer the direct questions: "Have you ever been in an institution; what's it like to hold the hand of someone you love; do you ever feel dread," K responds with hesitation, thus revealing his anxiety, but also responding to language in the form of "communication" rather than as neutral reiteration.[7]

If no predicate can successfully be applied to Woman, if she follows not the logic of the Law of exclusion, but the impasse of the 'not-all', then she is also free to enact the impossible. Her acts cannot be located within the system of the Law and its transgression, because they are singular and call into question the entire structure of the Law and transgression. And with this orientation to the feminine, let us take a closer look at what might have prompted Denis Villeneuve to say that his film shows many facets of femininity. Here we can readily see that while K searches for his origins, for his father, for his "authentic" childhood and memories, the women (all except for Luv), are doing something entirely different: Freysa and Mariette are leading an underground revolution and open up the space of the political; Joshi goes beyond her charge to maintain order and sacrifices herself in order to protect K; Ana is capable of creating histories that serve as the Real, the backdrop against which myths, identities, desire, and sex are all formed; Rachael (present mainly in her absence in this film) gives birth to a girl who becomes the symbol of replicant freedom; and Joi chooses to be with her lover, K, no matter the cost. Whereas K's quest is to find his lineage in time, a lineage which will also guarantee a kind of future history, the women act according to a necessity, which is paradoxically both located in space and time and universal, as that which refuses the limits of space and time. When Freysa tells K that no one's individual life means anything compared to the revolution that is to come, it is also an attempt to retain the choice to offer one's life in a sacrifice that simultaneously opens the space of a "cause" that is worth sacrificing for.

Although the world of *Blade Runner 2049* is one of diminishing appearances, each one of the women of the film is seen through her distinct appearance, her masquerade. Joi, in particular, is capable of projecting a different appearance instantaneously, according to K's mood, and this includes her dressing as a 1950s American housewife when he comes home from work, her changing into a mini-skirt and long hair to express her happiness when K presents her with a gift, and then the more casual black pants and top she wears when relaxing with K in his apartment. Perhaps the most surprising thing about Joi is that although she is designed to fulfill an owner's sexual fantasies and there are many "Joi's" in existence, we don't see her clad in the attire of a sex worker, like Mariette, and this is mainly because K's relationship with her is as a subject, and is not limited to her as a computer program or as mere pleasure object. In return, Joi understands the central importance of K's fantasy and his search for his authenticity. While K searches for clues of the miracle child at the DNA Machine, Joi whispers in his ear that he is special, "of woman born, pushed into the world, wanted, loved." She feeds his fantasy and points out the coincidence of the date stamped on the wooden figure and the one he found carved into the tree where he found Rachael's remains. Further, she gives him a name, "Joe," which is also the only male name she's programmed to say. She is made up of zeroes and ones, but she hints that she may be an artificial intelligence, capable of becoming something beyond her program. Her status as subject is confirmed, however, not only in her act of naming K, but also in her choice to sacrifice herself for him. She tells K to break the antenna that connects her emanator to the grid in his apartment, so she will be free to go with him on the run, and which ultimately results in her end, when Luv steps on and crushes the emanator in an instant.

The other women of the film also appear according to their symbolic roles, for example, Lieutenant Joshi has short slicked-back hair and wears a distinctly masculine uniform that appears to signify her as phallus, since she is, as head of the LAPD, in charge of keeping order, of maintaining the wall between humans and replicants, and of finding and killing replicants who have gone rogue against the Wallace corporation. But Joshi also gives hints that she is more than her symbolic role as Police lieutenant when she asks K to tell her a childhood memory, even though it might be implanted. This offers K an opening to share a crucial memory about his fantasy object, the carved wooden horse that he hid at the orphanage after being chased by other little boys. Joshi in the end assures K that he has done all right despite the the fact that he doesn't have a soul. Although Joshi tells K that if the knowledge of Rachael's baby would get out it would "break the world," she makes a crucial choice that reveals she is willing for this very "break" to occur. After K goes on the run and Luv comes to Joshi's office demanding to know where he is, Joshi ends up sacrificing her life to save K, and in this emerges the realization that the "wall" she spoke of earlier is not between replicants and humans but between those, like K and herself, acting to free the enslavement of all and the empire of theWallace Corporation. When while crushing Joshi's hand holding the glass, Luv tells her, "you can't hold the tide [of the Wallace empire] with a broom," Joshi replies, "except that I did."

The three other women, Freysa, Ana, and Mariette also appear according to their symbolic roles, but each one also acts to open the world to something beyond its present coordinates. Mariette is programmed as a Doxie sex worker, but her activity in the underground and her efforts to enlist K reveal she has a political cause, as does Freysa who is leader of the nascent underground movement. When Freysa meets K, she appears in a full-length robe wearing dark sunglasses to hide her missing right eye, which was removed so she cannot readily be identified as an older Nexus replicant. She has a history fighting on Calantha with Sapper Morten (Dave Bautista), holding Rachael in her arms as she died giving birth, and helping to hide the miracle child at the age of 8 in an orphanage. Freysa is the one who tells K that the ultimate human act is to "die for the right cause."

Dr. Ana, who must live encased in a glass room due to a compromised immune system, is the female who appears as the most casual, since she is unencumbered by the need to appear sexy or alluring. She has acquired the status of an erudite professional who has the unique ability to create memories for replicants, so that they will have a history, which further allows them their narrative and sense of connection. When K reveals his crucial memory of the wooden horse to her, she recognizes the memory as hers, when she herself was in the orphanage. She tells K with tears in her eyes that his memories are real, that the events really happened. And while she doesn't reveal that they are her real memories, she tells him that in her ability to give replicants memories they can also then have "real human responses." Although it remains unclear what type of being Ana herself is (whether she was born of two replicants or a replicant mother and human father), her abilities to provide memories for replicants turns out to be the key for them to organize their narratives that allows for the hysterical question about their purpose. Their castration into the discursive order of language is what ushers in their ontological uncertainty wherein they will become sexed subjects.[8] Memories and their ability to convey humanness have nothing to do with being "authentic" or detailed; rather, as Ana tells K, "They all think it's about more detail, but that's not how memory works; we recall with our feelings; anything real should be a mess."

The women of *Blade Runner 2049* take on various appearances, and which reveals that the essence of woman is indistinguishable from her masquerade. The feminine as masquerade stands in stark contrast to the way Wallace treats sex and the feminine as mainly a matter of biological engineering, as a way to ensure his empire of millions upon trillions into the future. It is the women, however, who identify what is singular, and in this regard Woman refers here not to a hidden core of the Real beneath symbolic reality, but rather to the crack in being, and in reality itself. By their acts (of necessity), the women of this film open the space for subjectivity, the space around which fantasies and myths appear, where notions of freedom, causality, and sacrifice are generated, where enjoyment (sexuality) and sex take shape. And here only Luv, who in her search to obtain the recognition of her uniqueness, reveals a sexuation configured within a masculine impasse. She is more like an adolescent boy whose emotions at times run amok while she aspires to provoke the recognition and approval of her father.[9] When K first meets her and finds that she has a name, he relates that since Wallace named her she must be special; Wallace himself calls her "special," and "the best angel of all"; and when she is about to kill Lieutenant Joshi she says, "I'm going to tell Mr. Wallace you tried to shoot me first," a blatant lie that attempts to justify the killing in the eyes of her father.

The ideological battle occurring in this film is the opposition between being (a being that retains its singularity) against the bioengineered replicants produced only for the expansion of Wallace's empire. What we detect here is a myth of the organic, but the organic re-framed; that is, it is more authentic (and desirable) to be born of replicants than it is to be bioengineered by Wallace, but this is not the key to what "being human" is all about. When K meets Freysa and the replicants of the underground forming a resistance, and it is revealed to him that he is a replicant manufactured by Wallace, and not the "miracle child" he had thought, Freysa says: "oh you imagined it was you; we all wish it was us." But she immediately specifies that there is something more important than being born of "real" parents and that is that there is no greater human act than to die for the right cause. And in this regard, Joshi, who dies fighting Luv without revealing where K is, Mariette and Freysa who are ready to sacrifice their lives for the revolution, and Joi who chooses to go with K knowing that it will no doubt mean she will be destroyed, are all "human," as is, in the end, K himself who dies for his "right cause," in rescuing Deckard and reuniting him with his daughter.

It is from a psychoanalytical perspective that we can see that within the world of *Blade Runner 2049*, a world of the loss of appearance itself, there is an attempt not to overcome sex (as the negative space that pertains to the discursive) but to fight against the disappearance of this gap, the absence around which identities take shape. This brings us, however, to a crucial question: why, in a future post-human world do we need sex and sexual difference at all; why not create a post-human world that envisions what current advocates of pan-sexualism, fluid gender identities and even asexuality, claim as a right of self-determination against the exclusive binary of masculine and feminine sexes of today's identity politics? The answer lies in what the ontological impasse of sex makes possible. As Žižek puts it, the growing list of classifications we find in LGBTQA+ (and in particular the addition of 'asexuality', as well as the plus sign at the end) means, paradoxically, that such a list itself signals "the persistence of sexual difference as real, as 'impossible' (defying every categorization)…"; what the ever-increasing list reveals is that sexual positions themselves circulate "around an antagonism that forever eludes it" (Zizek, 2017; 137). As Lacan revealed in his formulas of sexuation in *Seminar XX*, and as many astute theorists after him, and in particular Joan Copjec and Alenka Zupančič, have illustrated in so many brilliant ways, sexual difference is a different (and singular) kind of difference, one which does not follow the logic of binary oppositions, but rather occurs where logic itself breaks down, thus illuminating also our incomplete ontology as sexed subjects.[10]

In *Incontinence of the Void* Žižek refers to a "unique process" occurring today, one that speaks of "the gradual disintegration of our most basic sense of reality," which includes the effects of virtual realities, the passage to a trans- or post-human dimension, the loss of the binary form of sexuality, as well as a global capitalism that with its technologies is ushering in a "post-patriarchal, nihilist order of consumption" (Zizek, 2017; 140–141). As he posits, although we are losing the traditions on which we can base our identities, we can detect capitalism's attempt to fill in with what Lacan called *lathouses*, or the superabundance of objects designed to cause our desire, generated by corporate capitalism's scientific and technological intervention into the Real of our everyday lives, including cell phones, remote-controlled drones, gadgets of techno-sex, artificial organs,, etc. As Žižek proposes, *lathouses* work on the one hand to dispel the myth of One-Woman, but on the other, they enable the avoidance of seeking an answer to "the Woman question" itself; of seeing how Woman is the crack in reality, the gap around which positivized identities are formed (Zizek, 2017; 144). Ultimately, we can detect in the narrative and imageries of *Blade Runner 2049* this very "unique process" of which Žižek speaks; that is, we detect the tension between the mythical feminine essence (replicant Woman as the bearer of a miracle child), and pleasure administered increasingly by holographic projections, Doxie sex workers, and computers.

As argued in this film analysis, the two subject-positions of masculine and feminine are two different ways to address the symbolic dimension's inability to offer complete ontological consistency, or two ways to address the non-sexual relation. Subjects are not "constructed" by language; they are produced as a response to its inherent limit, and where a surplus appears. Or, as Zupančič put it: "Minus one/plus enjoyment – this is the necessarily distorted structural topology where the subject of the unconscious dwells" (Zupancic, 2017; 62). *Blade Runner 2049* goes to great lengths to make use of obviously exaggerated and traditionally patriarchal images of sexualized females while it presents a future that threatens appearance, and the possibility of sex in all its radical ambiguity. Not surprisingly, this impasse or "in-between" world is mirrored in Director Villeneuve's conscious effort not to rely too much on overly-synthesized or computerized sets like green screens or digitalization; rather, he worked hard with his team to produce three dimensional sets and maintain references to analogue technologies where possible. Thus we encounter references to things we'd expect would have become obsolete, for example, the DNA Machine that K uses reminds us of a microfiche machine in archival retrieval of the twentieth century. And Villeneuve further chose not to use variations of our current Iphones or readily recognizable data retrievable such as the cloud, but instead opted to create tangible objects, such as Ana's camera for creating memories, and the crystal orbs for archived data recording.

In *Blade Runner 2049* the sexualized imageries of women appear as vestiges of patriarchal fantasies that work to conceal the crucial role that the women of this film play. These women don't solve the impasse of the in-between world of the future and its losses, but rather, through their acts and sacrifices, they point to ways that another world might be created, one where the constitutive gap of the discursive order would allow for the dimension of sex (and subjectivation) itself. If, as Zupančič relates, patriarchal society and its oppression of women is a vast attempt to deal with and overcome the fact that Woman doesn't exist, this doesn't stop women from assuming all kinds of identities in the symbolic realm; they continue to be wives, mothers, daughters, lovers, etc. "This abundance of symbolic identities disguises the lack that generates them" (Zupancic, 2000; 132). But it is by acknowledging Zupančič's further claim that the loss of the feminine "is most unbearable for men, since it calls into question a portion of their own being, invested as it is in the symbolic roles of the woman," that allows us a way to analyze what I consider to be the weakest feature of this otherwise brilliant film, and that is the narrative's ending (Zupancic, 2000; 132–133). Although in her appearance and role Ana projects a casual, unassuming kindness, and she is a very accomplished woman who loves her work (having the title of "Dr."); she has no knowledge that her birth was a "miracle"; that her very being will provide the myth of the One-Woman that will become the symbol of replicant freedom and resistance. For Deckard, on the other hand, the meeting of his daughter for the first time will give him his symbolic mandate, his purpose of becoming her protector, shielding her from Wallace's evil designs, whether this means he will go back into hiding or go on the run with his daughter as he once did with her mother. The focus on this "couple" works to redirect the frame through which we see not only the ending of this narrative (the coming apocalyptic battle reduces down to a father–daughter relationship), but also provides the opening for the next installment of the *Blade Runner* series, which might be titled: "The Battle of Two Fathers." And it is primarily in this sense, and not mainly in the overly-emphasized sexualized female imageries of the film, that we can detect the male fantasy at work.

Notes

1. In Zupančič's intricate treatment of sex and sexuation in her phenomenal theoretical work, *What Is Sex?*, she reveals in various ways how sex and sexuation occur simultaneously with the emergence of the signifying order, which itself directly coincides with "the non-emergence of one signifier" as the place of a gap where enjoyment appears, "as the heterogeneous element pertaining to the signifying structure, yet irreducible to it." This enjoyment belongs to the unconscious and pertains to the drives, which relates to Lacan's formulas of sexuation and the "question of knowledge and its 'limits'" (53).
2. In Chapter 7 "A love letter (*une lettre d'ậmour*) Lacan offers his graph of sexuation with four formulas and throughout this chapter he will explain how the masculine is located at the impasse of a negation of exclusion (through the phallic function) and which leads to fantasy, the object *a*, and "the *jouissance* of the idiot." While the logic of the Woman rests on the following: "Woman has a relation with S(Ⱥ), and it is already in that respect that she is double, that she is not-whole, since she can also have a relation with ϕ," 81. *The Seminar of Jacques Lacan, On Feminine Sexuality, The Limits of Love and Knowledge, Book XX, Encore 1972–73*, Ed. Jacques-Alain Miller, Tr. Bruce Fink (New York: W. W. Norton, 1975), 80–81. According to Joan Copjec, Woman's existence can neither be contradicted nor confirmed by reason, and this is why Lacan states the "Woman does not exist." She writes that Lacan "leaves open the possibility of there being something – a feminine *jouissance* – that is unlocatable in experience that cannot, then, be said to exist in the symbolic order." Copjec, *Read My Desire*, 224.
3. Neroni writes: "When feminist film theory focuses solely on identification, it misses the failure of identity and the failure of ideology. And by doing so, it paints a picture of ideology that is itself ideological because it offers precious little room for challenging ideology" (35).

4. This perspective is articulated by Slavj Žižek in his analysis of the deadlock of sexual classifications and the shift to a post-human world, which amounts to "at its most fundamental…the overcoming (leaving behind) of the sexual in its most radical ontological dimension – not just 'sexuality' as a specific sphere of human existence but the Sexual as an antagonism, the bar of an impossibility, constitutive of being-human in its finitude" (Žižek, 2017; 134).
5. As Žižek explains it, there is an asymmetry in sexual difference which resides in a slit or cut that is different for each impasse: For Man "the macho-image is not experienced as a delusive masquerade but as the ideal-ego one is striving to become. Behind the macho-image of a man there is no secret, just a weak ordinary person that can never live up to his ideal; whereas the trick of the feminine masquerade is to present itself as a mask that conceals the feminine secret [that doesn't exist]." Slavoj Žižek, "Woman is One of the Names-of-the-Father: Or, How Not to Misread Lacan's Formulas of Sexuation," *lacanian ink* 10, 1995. And this is also how Žižek comes to ultimately conclude: "If Woman does not exist, then man is Woman pretending to exist" Žižek, 2014; 45).
6. K's repeating of lines from Vladimir Nabokov's *Pale Fire* can also be taken to reveal the existential anxiety K has about his being, as a replicant shunned by humans, searching for his father, and destined for a life without authenticity.
7. This prompts us to consider Wallace's use of language, which conveys a kind of God-like, egomaniacal rhetoric, where he refers to his corporation as "the Kingdom of Heaven," and the future as an "Eden" he will create with the enslavement of "millions so that there can be trillions" more replicants to colonize even more worlds. But in this raving we can detect an ego-ideal at work; there is no radical ambiguity in how he envisions himself as Father-creator.
8. The nature of Ana's being depends on one's preference of the ending of either the original film (released in 1982) or the Director's Cut released in 2007; in the latter (and as Ridley Scott has specified) Deckard is a replicant, but it is left ambiguous in *Blade Runner 2049*. This comes across when Deckard is taken to Wallace and the latter asks Deckard to consider whether he wasn't specifically designed for the purpose of meeting and falling in love with Rachael so as to procreate.
9. In an interview in an Extra of *Blade Runner 2049* "Designing the World of Blade Runner Special Effects," actress Sylvia Hoeks relates that both she and the Director conceived of her character Luv as more like a 12 year

old, sometimes acting like a boy, sometimes like a girl, and other times as a woman.

10. As Zupančič writes, "sex is messy because it appears at the point of the breaking down of the signifying consistency, or logic (its point of impossibility), not because it is in itself illogical and messy: its messiness is the result of the attempt to invent a logic at the very point of the impasse of such logic" (Zupancic, 2017; 43).

References

Copjec, J. (2002) *Imagine There's No Woman.* Cambridge: The MIT Press.

Copjec, J. (2015) *Read My Desire: Lacan Against the Historicists.* London: Verso.

Fletcher, R. and Ashurst, S. (2018) "Can We Talk About 'Blade Runner 2049's' Problem with Women?," *Digital Spy*, accessed January 2018, available at: http://www.digitalspy.com/movies/feature/a839916/bladerunner-2049-gender-issue.

Friedlander, J. (2008) *Feminine Look: Sexuation, Spectatorship, Subversion.* New York: SUNY Press.

McGowan, T. (2007) *The Real Gaze: FilmTheory After Lacan.* New York: SUNY Press.

Neroni, H. (1996) *Feminist Film Theory and 'Cleo from 5 to 7'.* New York: Bloomsbury.

Smith, A. (2017) "Is *Blade Runner 2049* Sexist – or a Fair Depiction of a Dystopian Future?" *The Guardian*, October 9, available at: https://www.theguardian.com/film/2017/oct/09/is-blade-runner-2049-a-sexist-film-or-a-fair-depiction-of-a-dystopic-future.

Žižek, S. (1993) *Tarrying with the Negative: Kant, Hegel, and the Critique of Ideology.* Durham. Duke University Press.

Žižek, S. (2014) *The Most Sublime Hysteric: Hegel with Lacan.* Cambridge: Polity Press.

Žižek, S. (2017) *Incontinence of the Void: Economico-Philosophical Spandrels.* Cambridge: The MIT Press.

Zupančič, A. (2000) *Ethics of the Real: Kant, Lacan.* London: Verso.

Zupančič, A. (2017) *What Is Sex?* Cambridge: The MIT Press.

Index

C. Neill (ed.), *Lacanian Perspectives on Blade Runner 2049*, The Palgrave Lacan Series,
https://doi.org/10.1007/978-3-030-56754-5

The manufacturer's authorised representative in the EU is Springer Nature Customer Service Centre GmbH, Europaplatz 3, 69115 Heidelberg, Germany. If you have any concerns regarding our products, please contact ProductSafety@springernature.com

Printed and bound by CPI Group (UK) Ltd, Croydon, CR0 4YY
30/09/2026
02225451-0004